Introduction to Animal Science

Second Edition

Rick Parker • Karen Kenny • Miriah Pace

© 2017 Second Edition - National Agricultural Institute, Inc.

ALL RIGHTS RESERVED. No part of this work covered by the copyright herein may be reproduced except as permitted under Section 107 or 108 of the United States Copyright Act, without the prior written permission of the publisher.

Published by:

National Agricultural Institute, Inc.
151 W 100 S
Rupert, ID 83350
USA
(208) 957-7000

ISBN 978-1-4951-5851-3

To learn more about the National Agricultural Institute visit:
www.national-ag-institute.org

Or visit us on Facebook: www.facebook.com/NationalAgInstitute/

Send comments or questions to: info@national-ag-institute.org

Thanks to the feedback from the first edition, our second edition has been revised. Minor errors and broken links were corrected as well as the addition of more illustrations to create a more effective teaching tool.

Notice to the Reader

Publisher does not warrant or guarantee any of the products described herein or perform any independent analysis in connection with any of the product information contained herein. Publisher does not assume, and expressly disclaims, any obligation to obtain and include information other than that provided to it by the manufacturer. The reader is expressly warned to consider and adopt all safety precautions that might be indicated by the activities described herein and to avoid all potential hazards. By following the instructions contained herein, the reader willingly assumes all risks in connection with such instructions. The publisher makes no representations or warranties of any kind, including but not limited to, the warranties of fitness for particular purpose or merchantability, nor are any such representations implied with respect to the material set forth herein, and the publisher takes no responsibility with respect to such material. The publisher shall not be reliable for any special, consequential, or exemplary damages resulting, in whole or part, from the readers' use of, or reliance upon, this material.

Table of Contents

1 Animal Domestication 1
- Primary and Secondary Energy Traps 1
- Human Evolution 1
- Symbiosis 2
- Domestic Animal 2
- When, Where, Why to Domesticate 3
- Domestication of Modern Species 3
- Importance of Animal Domestication 4
- Timeline of Domesticated Animals 4

2 Nomenclature 6
- Nomenclature 6
- Classification 6
- Challenge of Classifying Living Things 7
- Early Classification of Living Things 7
- Modern Classification 7
- Five Kingdoms 8
- Binomial Nomenclature 9
- How it Works for Domestic Animals 9

3 Species 11
- Species 11
- Subspecies 12
- Breed 12
- Subspecies, Breed, Race and Variety 12
- Scientific Naming of Animals 13
- Scientific Names for Farm Animals 13

4 Animal Protein 16
- Meat and Protein 16
- Animals as a Source of Food 17

Table of Contents

    Meats 17
    Other Protein Sources 18
    Basic Animal Science Enterprises to Produce Protein 19
    U.S. Meat Animal Production in Billions of Pounds 20
    U.S. Milk Production 20
    U.S. Poultry and Egg Production 21
    Value of Livestock Production to U.S 21
    Aquaculture 21

5 Meats 24
    Carcass Grades 24
    Dressing Percent 27
    Wholesale and Retail Cuts 28
    Wholesale and Retail Cuts of Beef 28
    Wholesale and Retail Cuts of Pork 29
    Wholesale and Retail Cuts of Lamb 29

6 Hormones and Meat Production 31
    What is a Hormone? 31
    What is a Steroid? 32
    Synthetic Steroids Used in Meat Production 32
    Misuse of Hormones 34

7 Milk and Eggs 36
    Milk Production 37
    Major Sources of World Milk Production 37
    Milk Production Steps 37
    Milk Grade 38
    Nutritive Value of Milk 39
    Milk Fat 39
    Carbohydrates 39
    Proteins 40
    Vitamins 40
    Minerals 40

Table of Contents

   Homogenization 40
   Pasteurization 41
   Producing Fluid Milk 42
   Classified Pricing 42
   Eggs 43
   Nutritive Value of Eggs 43
   Steps in the Processing of Eggs 43
   Size Determination 45
   Packaging of Eggs 45
   Care and Handling of Eggs 46
   Labeling 46

8 Wool 49
   Value of Hair/Wool from Mammals 49
   Growth of Wool, Hair and Mohair 50
   Factors Affecting the Value of Wool 51
   National Wool Act and Incentive Payment 51
   Classes and Grades of Wool 51
   Uses of Wool and Mohair 52
   Qualities of Wool 53
   Wool Production in the United States 53

9 Animal Cells 55
   Three General Functions of Most Cells 55
   General Cell Structure 56
   Organelles Unique in Animal Cells 58
   Organelles Unique in Plant Cells 58
   Types of Cells 58

10 Cell Functions and Types 60
   What is a Cell? 60
   Functions of Cells 61
   Types of Cells 61
   Blood Cells 62

Table of Contents

    Absorptive Cells 63
    Secretory Cells 63
    Nerve Cells 64
    Sensory Cells 64
    Muscle Cells 65
    Reproductive Cells 65
    Bone Cells 66
    Fat Cells 66

11 Mitosis and Meiosis 68
    Mitosis 68
    Meiosis 71

12 Introduction to Anatomy/Physiology 74
    General Information 74
    Some General Anatomical Terms and Understandings 75
    Body Systems 77

13 Comparative Anatomy 81
    Introduction 81
    Adaptation, Comparative Anatomy, Homology and Analogy 82
    In Contrast to Homologous is Analogous 82

14 External Anatomy of Farm Animals 84
    Parts of a Dairy Cow 85
    Parts of a Horse 86
    Parts of a Pig 87
    Parts of a Goat 88
    Parts of a Sheep 89
    Parts of a Chicken 90
    Parts of a Beef Animal 91

15 Skeletal System 93
    Bone Structure 93
    Skeletal System 94
    Axial Skeleton 94

Table of Contents

    Appendicular Skeleton 96

    Articulation of the Joints 97

16 Digestive System 99
    System Components 99

17 Hormone Function 105
    Endocrinology 105

    Location of Endocrine Glands 106

    Endocrine Glands, Hormones and Functions 107

    Study of Hormone Function in Males 108

18 Hormone Control 110
    Hormones Control and Influence 110

    Feedback Control 113

19 Reproductive Systems 115
    Overview of Animal Reproduction 115

    Male Reproductive Tract 116

    Female Reproductive Tract 119

    Reproduction in Males 120

    Reproduction in Females 120

    Estrus and the Estrous Cycle 121

20 Fertilization 124
    Gametogenesis 124

    Spermatogenesis 125

    Oogenesis 125

    Oocyte 126

    Fertilization 126

    Female Gamete: The Ovum 126

    Male Gamete: The Sperm 127

    Fertilization Process 127

    Visual Representation of Gametogenesis, Chromosome Count and Fertilization ... 128

    Segregation 128

Table of Contents

21 Conception, Gestation and Parturition 130
Conception, Sex Cells and Fertilization 130
Placenta 132
Gestation 132
Parturition 133

22 Inheritance 135
Brief History of Genetics 135
Chromosomes and DNA 136
DNA: Deoxyribonucleic Acid 137
Genes 138
Genetic Codes and Protein Synthesis 138

23 Interaction of Dominant and Recessive Genes 141
Characteristics of Genes 141
One Pair of Genes 143
Two Pairs of Genes 144
Sex Linked Inheritance 145
Heredity vs Environment 145

24 Understanding Genotype and Phenotype 147
Introduction 147
Punnett Square 148
Pea Plants Characteristic to Study: Stem Length 148

25 Selection 151
What is Evolution? 151
Genetic Variability 152
Natural Selection 152
Use of Genetic Variability and Natural Selection in Producing Improved Livestock. 153

26 Mutation 155
Mutation Defined 155
Examples of Mutations Useful to Commercial Agriculture 156
Human-Induced Mutation 157

Table of Contents

27   Digestion and Absorption 159
Digestive Systems 159
Ruminant Defined 160
Non-ruminant (Monogastric) Defined 160
Four Compartments of the Ruminant 160
Functions of the Compartments 160
Non-ruminant Digestion 162
Understanding Digestion and Absorption 162

28   Nutrient Needs 165
Nutrient Needs Definitions 165
Six General Classes of Nutrients (Needed by all livestock and humans): 166

29   Protein Needs 170
Protein Needs 170
Excess Protein 170
Essential Amino Acids 171
Amino Acid Requirements of Animals 171
Protein in the Rations of Ruminants and Non-Ruminants: 171

30   Carbohydrates and Fats 174
Carbohydrates Needed in Livestock Feed 174
Fats in Livestock Feeding 175

31   Vitamins and Minerals 178
Vitamins and Their Functions in Livestock Feeding 178
Function, Deficiency Signs and Sources 179
Minerals and Their Functions in Livestock Feeding 183
Function, Deficiency Signs and Sources of Minerals 183

32   Feed Composition 188
Analysis 188
Energy Terms and Calorie 190
Feed Classifications 191
Importance of Feed Composition 191

Table of Contents

33 Feed Preparation 193
Feed Preparation Methods 193

34 Feed Additives 197
Feed Additives Defined... 197
Feed Additives 198
Feed Additives Sources 198

35 Computing Balanced and Least Cost Rations 200
Livestock Rations... 200
General Guidelines 201
Square Method (called Pearson Square): 201
Feeding Standards Method 202
Least Cost Balanced Rations 204
Guides for Concentrates and Roughages.. 205
Computing the Cost of a Ration 206
The Art and Science of Feeding 206

36 Lactation... 208
Lactation Defined 208
Mammary Gland Structure... 209
Species Differences... 209
Mammary Gland Development and Milk Secretion.......................... 209
Maintenance of Lactation 210
Factors Affecting Milk Production 210
Nutrients 210
Using Dairy Cow as an Example 210
Importance of Feeding Colostrum to Calves................................... 211
Tracing Milk from the Udder to the Milker 212
Milk Letdown.. 212

37 Factors Affecting Animal Health 214
Preventative Herd Health Management Programs 214

38 Animal Health Evaluation... 218
Animal Health Observed.. 218

Table of Contents

Physical Examinations 219

39 Responsibility for Animal Health 224
Animal Health Concerns 224
Animal Owner 224
Veterinarian 225
Government 225

40 Disease Causing Agents 228
Disease 228
Infectious Agents Causing Diseases 228
Life Cycles of Some Common Parasites 230
Fungi 232
Non-Infectious Agents that Cause Disease 233

41 Development and Types of Immunity 235
Healthy Body against a World of Disease 235
Immunity 236
Immunity or Resistance to Disease – Two Groups 237
Disease Development 238
Ways Diseases Spread 238
Good Management For Disease Prevention 239

42 Vaccination and Administration of Biologic Agents 242
Vaccination 242
Administering Biologic Agents 244
Methods of Treatment Application 244
Methods of Injection 245

43 Beef Cattle Industry 247
International 247
In the United States 248
Type of Beef Cattle Enterprises 248
Beef Cattle Breeds 250
Cow-Calf Management 250
Breeding 251

Table of Contents

Artificial Insemination (AI) 252

Crossbreeding 253

Replacement Heifers 253

Brood Cow Characteristics 253

Bulls 254

Calving Season Management 254

Calf Management 254

Summer Management: Cow-Calf Enterprise 256

Common Forages for Beef Cows 257

Pasture Management 257

Rules of Thumb for Winter Feeding Breeding and Stocker Cattle 258

44 Dairy Cattle Industry 261

Place of Dairy Cattle in U.S. Agriculture 261

Purpose of Dairy Industry in the U.S. 262

Historical Perspective 262

Structure and Geography 262

Trends in Dairies 263

Herd Improvement 263

Dairy Selection and Breeding Program 264

Dairy Breeds 265

Reproductive Management 265

Artificial Insemination 266

Nutrition in Dairy Cattle 266

Dairy Herd Health 266

Bovine Somatotropin (BST) 267

Trends and Influential Factors for the Future of Dairying 267

Food Safety in the Dairy Industry 268

Environmental Concerns for Dairies 268

45 Swine Production Industry 270

Common Breeds of Swine in the U.S. 270

Husbandry Systems 270

Table of Contents

    Farrowing Systems 271
    Litter Management 271
    Nursery Systems 272
    Growing and Finishing Systems 272
    Mating Systems 272
    Sow Management during Gestation 272
    Good Management Practices 273
    Handling 273
    Transportation 274
    Environmental Management 274
    Facilities and Equipment 275
    Feeding and Nutrition 277
    Balanced Diet Considerations 278
    Feeding Practice Groupings 278
    Herd Health Management Concerns 278
    Surveillance, Diagnosis, Treatment and Control of Disease 278
    Pork Quality Assurance Plus 279

46 Sheep 281
    Characteristics of Sheep 281
    Predominant Breeds 282
    Management of Sheep 282
    Potential 283
    Production 284
    Future Needs 284
    Management Needs 285
    Nutrition of Sheep 285
    Lamb Management Activities 286
    Components of Health Program 286
    Facilities 287
    Wool 287

Table of Contents

47 Goats 289
Overview of Goats 289
Predominant Breeds/Types 290
Management of Goats 291
Feeding and Health Program 292
Production 293

48 Horses 296
Breed Definition and Color 296
Gaits 297
Major Horse Categories, Breeds and Characteristics 298
Uses of Horses 300
Selecting and Judging Horses 300
Parts of a Horse and Their Ideal 301
Conformation 303
Feeding Horses 305
Feeding Guidelines 306
Signs of Good Health 308
Horse Health Program 309
Tack 310
Basic Horsemanship 310
Horse Shows and Competition Include 310

49 Poultry 312
Vertically Integrated Structure of the Modern Poultry Industry 312
Value to the Economy 315
Common Breeds of Meat and Laying Chickens 315
Specific Breeds Used in Industry 316
Common Meat Breed Used in the Commercial Turkey Industry 317
Processing Eggs 317
Poultry Processing 317
Further Processing 320
Storage 320

Table of Contents

Requirements of a Packaging Label 320

50 Aquaculture 323
Aquatic Species 323
Characteristics of Fish 324
Dietary Protein Needs 324
Dietary Carbohydrate Needs 325
Dietary Lipid Needs 325
Dietary Energy Needs 325
Vitamin Needs 325
Mineral Needs 326
Toxins 326
Feeding Programs 326
Feed Formulation and Processing 326
Feeding Methods 327
Typical Diet Components 327
Aquaculture Activities 327

51 Companion Animals 329
Place of Pet and Companion Species in U.S. 329
Who Owns Pets? 330
How Companion Animals Connect to Agriculture 330
Pet and Companion Animals Defined 331
Historical Perspective of Dogs and Cats 332
Genetics and Breeding Programs 332
Development of Breeds 333
Breeds of Dogs 333
Breeds of Cats 333
Reproductive Characteristics of Various Pets/Small Animals 334
Nutrition of the Pet Species 336
Trends in Veterinary Expenditures 336

52 Animal Behavior 338
Definitions of Animal Behaviors 338

Table of Contents

    Anthropomorphism 340

53 Issues in Animal Science 342

    Beef Issues.. 342

    Dairy Issues 343

    Horse Issues 343

    Poultry Issues 343

    Sheep Issues 343

    Swine Issues 343

    Animal "Protection" Groups 344

    Animal Welfare Concerns: Beef.. 344

    Animal Welfare Concerns: Swine 344

    Animal Welfare Concerns: Dairy 344

    Animal Welfare Concerns: Veal 345

    Animal Welfare Concerns: Poultry 345

    Animal Welfare Concerns: Horses 345

    Model for Livestock Producers from Animal Protection Groups........................... 345

    Animal Welfare Act 346

    Biotechnology 346

    Public Impressions of Genetic Engineering 346

    Biotechnology Issues 347

    Environmental Issues 347

    Environmentalism 347

    Livestock Environmental Issues 348

    Waste Management... 348

    Water Use and Quality... 348

    Federal Lands 349

    Endangered Species 349

    Global Warming/Climate Change 349

    Listing of Consumer Issues/Concerns 349

    Diet and Health 350

    Food Safety Concerns 350

Table of Contents

Food Safety Risks in Perspective 350
Marketing Issues 351
Issues and Opportunities 351
Consumption and Consumer Attitudes 351
Appendix 354
Conversion Tables for Common Weights and Measures 354
Conversion Factors for English and Metric Measurements 355
National Agriculture, Food and Natural Resources (AFNR) Career Cluster Content Standards 356
    Purpose 356
    Process 356
    Alignments and Crosswalks 356
    Availabiltiy 357

Glossary 358

Preface

This textbook, Introduction to Animal Science, is one in a series of *Just the Facts* (JtF) textbooks created by the National Agricultural Institute, representing a bold, new approach to textbooks. These textbooks present the essential knowledge in outline format. Essential knowledge is supported by a main concept, learning objectives, connection to national agriculture, food and natural resource (AFNR) standards and key terms, at the beginning of each section. Content of the books is further enhanced for student learning by connecting with complementary PowerPoint presentations and websites through QR codes (scanned by smart phones or tablets) or URLs. Instructors and students will find the extensive table of contents useful for planning class presentations and getting to important topics. Each textbook is available in print and electronic formats.

The time is now for a new mindset about textbooks. Textbooks for the future need to take advantage of both print and digital technology, while keeping costs down.

Just the Facts series of textbooks provide a synergistic textbook model - print and digital working together to be better than either one alone. Moreover, in a time of increasing costs for textbooks, print copies of "*Just the Facts*" textbooks are priced substantially less than traditional textbooks.

The first of these new textbooks include:

- *Just the Facts: Introduction to Agriculture*
- *Just the Facts: Introduction to Biology*
- *Just the Facts: Introduction to Food Systems Science*
- *Just the Facts: Introduction to Plant Science*
- *Just the Facts: Introduction to Soil Science*

Other titles scheduled for release as part of the *Just the Facts* textbook series include: *Introduction to Agribusiness, Introduction to Sustainable Agriculture, Introduction to Aquaculture Science* and *Introduction to Equine Science*. More are in the planning stages.

Instructors using one of the textbooks for a class of 15 or more students are eligible to receive the complementary PowerPoint presentations, laboratory activities and final assessments.

Just the Facts textbooks are a project of the National Agricultural Institute created, written and assembled by:

Rick Parker, PhD, President
Marilyn Parker, Vice President
Karen Earwood-Kenny, Board Member & Assistant Editor
Miriah Pace, Board Member & Assistant Editor

1 Animal Domestication

Major Concept
Domestication of animals benefit societies and civilization.

Objective
- Describe the importance of animal domestication to humans

Key Terms
- Feral
- Mutualism
- Primary energy trap
- Secondary energy trap
- Symbiosis
- Taming

Chapter Resource

 Complementary *full color* illustrations, photos, charts and graphs are available by scanning this QR code or by following this URL: http://www.tagmydoc.com/AS01 This digital resource will enhance your understanding of the chapter concepts.

Primary and Secondary Energy Traps

- Every organism is a **primary energy trap** (not related to the food chain). To be alive is to be a primary energy trap.

- **Secondary energy trap** is what an organism uses to help it obtain food (camouflage, stick, gun, etc.) or to help it conserve energy (shelter, fur, etc.).

Human Evolution

- Oldest population lived approximately 2 million years ago.

- First evidence of a domestic animal, the dog, was 12,000 to 14,000 years ago.

- First domestic food animal was the sheep, less than 11,000 years ago.

- Cultivation of plants began approximately 9,000 years ago.

- Humans survived for about 99% of their known history without domesticated animals or cultivated plants.

Symbiosis

- **Symbiosis** is a biological situation in which at least two different kinds of organisms interact; these can include plants, animals, or plant and animal.

- Specific type of symbiosis between man and animal is termed **mutualism**. In this situation, the two species benefit from each other.

- Domestication of animals is an example of a symbiotic relationship. Man provides food and shelter to the animals, and they provide meat, milk and fiber for man.

    - Because of this mutualistic symbiotic relationship, both humans and domestic animals are secondary energy traps for the others; humans protect, feed and care for animals while animals provide draft power, meat, skins, milk, fiber, etc.

Domestic Animal

- Includes animals whose breeding is or can be controlled by humans.

    - Domestic animals include horses, goats, sheep and pigs.

    - Excludes zoo and circus animals, various rodents and primates, research animals and animals caught wild and tamed.

    - Includes reindeer, although they live free.

        ✓ People of northern Eurasia followed herds to stay alive and intervened by selective killing, taking milk, using sleigh animals and castrating older males to lessen fighting among them.

- **Taming** is on the path to domestication, but a tamed animal is not a domestic animal.

    - Animals caught in the wild and tamed are not considered domesticated. Once they are bred in captivity and selected for qualities, such as docility, size, color, horn size, meat production, etc., then they can be considered domesticated.

- A domestic animal, or one descended from a domestic population, cannot truly revert to being a wild animal.

- Domestic animals that return to nature and breed are termed **feral**, examples: wild horses and pigs.

When, Where, Why to Domesticate

- Hunters and gatherers began domesticating animals through taming, but without any purposes other than for things they already knew – meat, skins and bones.

- The end of the glacial period marked the beginning of domestication and people were transformed from hunters and gatherers to herders and farmers.

- People began harvesting and storing wild grasses (wheat and barley) that were plentiful. At this point in history, people could not leave for long periods of time for fear someone else would come along and damage or destroy their food sources. Consequently, the culture changed as small villages began to emerge.

- Women no longer had to spend as much time gathering food, which caused an increase in birth rate and decrease in infant mortality.

Chapter Resource

  - Men now had to hunt near the villages but more animals needed to be domesticated for personal food supply.

  - Through long experience and a decrease in the need for a nomadic lifestyle, many secondary uses of animals were realized such as milk, wool, power, war, sport and prestige.

Domestication of Modern Species

- The dog was first, about 12,000 to 14,000 years ago in Northern Iraq. (A skeleton of a wolf pup or dog was found buried beside a woman, under her hand, indicating a tamed wolf or a domestic dog).

- Livestock

  - Iraq – sheep about 11,000 years ago

  - Asia Minor – swine and sheep about 9,000 years ago

  - Greece – cattle, swine, sheep, and goats about 7,000 years ago

  - Central Asia – horses and donkeys about 7,000 years ago

- Pakistan – poultry about 5,000 years ago
- Egypt – cats about 3,500 years ago

Importance of Animal Domestication

- Protection
- Consistent source of protein
- Power
- Milk
- Transportation
- Shelter
- Fiber
- Higher standard of living
- More time to develop government, religious and creative pursuits

Timeline of Domesticated Animals

- 2 million yr – First human populations
- 30,000 yr – Modern human
- 14,000 yr – Glacial melt back and dog
- 11,000 yr – Recent age
- 10,000 yr – Sheep
- 9,000 yr – Swine and crops
- 7,000 yr – Cattle, goats and horses
- 5,000 yr – Poultry
- 3,500 yr – Cat

Summary

Every organism is a primary energy trap. To be alive is to be a primary energy trap. Secondary energy trap is what an organism uses to help it obtain food or to help it conserve energy. The oldest population lived approximately 2 million years ago. Domestication of animals is an example of a symbiotic relationship. Man provides food and shelter to the animals and they provide meat, milk and fiber for man. Definition of a domestic animal includes those animals whose breeding is or can be controlled by humans. Hunters and gatherers began domesticating animals through taming, but without any purposes other than for things they already knew – meat, skins and bones. The dog was first, about 12,000 to 14,000 years ago in Northern Iraq.

Additional Resources

HistoryWorld, History of the domestication of animals
http://tinyurl.com/nwl7w4l

Leinhard, J.H., Domesticating animals
http://www.uh.edu/engines/epi1499.htm

Assessment

Take the online assessment here: http://tinyurl.com/ansci-q1
Download and print the assessment by scanning this QR code or by going to this URL: http://www.tagmydoc.com/Ch01AnSci

2 Nomenclature

Major Concept

All animals and plants are classified, arranged into groups and named based on their natural relationships.

Objectives

- Explain briefly why living things are classified
- List the seven divisions in modern classification
- Name the five kingdoms
- Define binomial nomenclature and give one example

Key Terms

- Class
- Hierarchical order
- Invertebrates
- Kingdom
- Nomenclature
- Phylum
- Taxonomy
- Vertebrates

Chapter Resource

Complementary *full color* illustrations, photos, charts and graphs are available by scanning this QR code or by following this URL: http://www.tagmydoc.com/AS02 This digital resource will enhance your understanding of the chapter concepts.

Nomenclature

- Giving and using of names

- Names are given for something considered important and necessary to remember and use again.

Classification

- Grouping and arrangement of organisms into a **hierarchical order.**

- An important aspect of classifications is their predictive value; for example, if a characteristic is found in one member of a group of plants, then it is also likely to be found in the other members of that group.

Challenge of Classifying Living Things

- Classifying living things is not simple because of the diversity and difficulty in finding, obtaining and studying.

Early Classification of Living Things

- A Greek scientist named Aristotle (384-322 B.C.) was one of the first scientists who attempted to classify living things.

    o All living things were placed into two main groups or kingdoms: plants and animals.

    o Using these main groups of plants and animals, further classification was used:

        ✓ Living things that are green and do not move.

        ✓ Living things that are not green and move.

Modern Classification

- Scientists used Aristotle's classification for hundreds of years.

- With new discoveries, Aristotle's system became less helpful because more plants and animals were being found.

- In the eighteenth century, a scientist by the name of Linnaeus reworked the system of classification.

    o He used some of Aristotle's early ideas but added many of his own.

    o He first divided all living things into two main groups as Aristotle had.

        ✓ He called these groups Kingdoms.

- A **Kingdom** is the first and largest division of living things – plants and animals.

    o Linnaeus then continued to subdivide the kingdoms.

Chapter Resource

- Each new group within a Kingdom is called a **phylum**.
  - Size is not used in this grouping. Very specific traits based on appearance of certain parts are used.
  - Phyla (plural of phylum) are divided into even smaller groups.
    - ✓ These new groups are called classes.
- A **class** is the largest division of a phylum. These are again divided into smaller groups.
- Seven main divisions in the modern classification system, from largest to smallest:

  1. Kingdom
  2. Phylum
  3. Class
  4. Order
  5. Family
  6. Genus
  7. Species

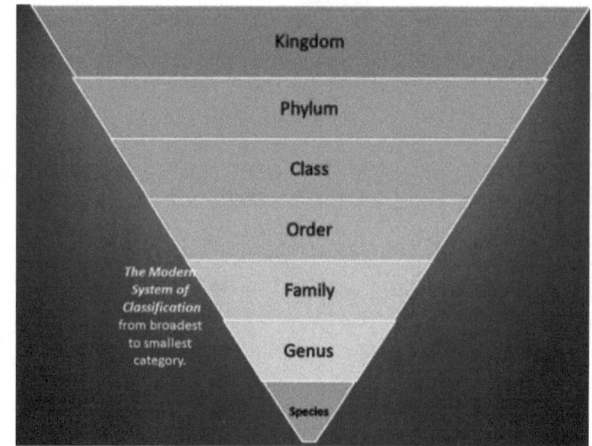

- Science of classification and the arrangement of plants and animals into groups based on their natural relationships are called **taxonomy**.

Five Kingdoms

- Over the last century, the kingdoms were expanded to five because it was difficult to group some living things into either plant or animal.

  1. Animal kingdom (Animalia) includes all living and extinct animals.
     - **Invertebrates** – animals lacking a backbone.
     - **Vertebrates** – animals with a backbone.

  Chapter Resource

  2. Plant kingdom (Plantae) includes all living and extinct plants.

  3. Fungi kingdom includes simple plants which lack chlorophyll, such as mold, mildew, rust and mushrooms.

JTF Introduction to Animal Science 8

- Many exist as parasites and saprophytes (which live on dead organic matter).
- Many are pathogenic (e.g., athlete's foot).

4. Monera kingdom includes bacteria and blue-green algae, characterized by the lack of true nuclei and chromosomes.

5. Protista kingdom consists of unicellular organisms with a true nucleus and chromosomes. All algae except blue-green algae belong to this kingdom.

Binomial Nomenclature

- Binomial nomenclature is a system of naming organisms also established by Linnaeus about 1758.
  - Each species is given a scientific name consisting of two Latin words:
    - ✓ The first, designating genus is capitalized.
    - ✓ The second, designating a subdivision of the genus, is not capitalized.
    - ✓ Normally both words are italicized, e.g. *Bos taurus* (European cattle)

Chapter Resource

How it Works for Domestic Animals

- Using cattle as an example:
  - Kingdom: Animalia (animals collectively)
  - Phylum: Chordata (animals with a backbone)
  - Class: Mammalia (warm-blooded, live young suckled by mother, etc.)
  - Order: Artiodactyla (even-toed, hoofed mammals)
  - Family: Bovidae (ruminants with poly-cotyledonary placenta)
  - Genus: Bos (ruminant quadrupeds with horns, etc.)
  - Species: *Bos taurus* (European) and *Bos indicus* (humped cattle)

Summary

A Greek scientist named Aristotle (384-322 B.C.) was one of the first scientists who classified living things. He placed all living things into two main groups, plants and animals. Later, another scientist, Linnaeus, created the basis of the system used today. All living things are classified into one of five kingdoms: Animal, (Animalia), Plant (Plantae), Fungi, Monera and Protista. A kingdom is the first and largest division of living things then, Phylum, Class, Order, Family, Genus and Species. The science of classification and the arrangement of plants and animals into groups based on their natural relationships are called taxonomy.

Additional Resources

Carl Linnaeus – Swedish Botanist
http://www.ucmp.berkeley.edu/history/linnaeus.html

Taxonomy: Life's filing system-Khan Academy
http://tinyurl.com/mv4y92d

Assessment

Take the assessment online here: http://tinyurl.com/AnSci-Nomenclature
Download and print the assessment by scanning this QR code or by going to this URL: http://www.tagmydoc.com/Ch02AnSci

3 Species

Major Concept

Naming systems and scientific names identify common production animals.

Objectives

- Define species, subspecies and variety
- Identify common livestock by their scientific name
- List two breeds for each of the following animals: horses, cattle, sheep and pigs

Key Terms

- Allopatric species
- Binomial nomenclature
- Breed
- Race
- Scientific name
- Subspecies
- Sympatric species
- Variety

Chapter Resource

 Complementary *full color* illustrations, photos, charts and graphs are available by scanning this QR code or by following this URL: http://www.tagmydoc.com/AS03 This digital resource will enhance your understanding of the chapter concepts.

Species

- A basic group forming a division of genus

    o Made up of a group of animals that closely resemble each other in body and physiology.

    o Can interbreed in a natural habitat and produce fertile offspring.

- Related species which cannot interbreed because of geographical separation are called **allopatric species**.

- Those which can interbreed, but in practice do not because of differences in behavior, breeding, food sources, etc., are called **sympatric species**.

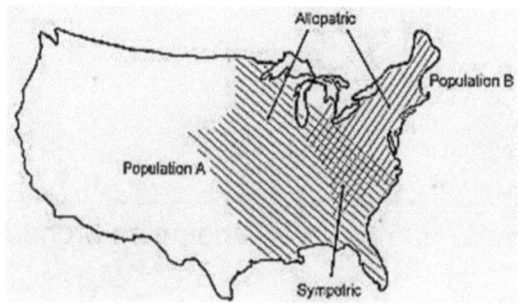

Subspecies

- A subdivision or smaller part of a group of animals (those in a species).

  o Usually a geographic race.

  o Differ from the other subspecies or groups of the same species in appearance or more rarely behavior.

Breed

- A race or variety of an animal species, especially one developed through human influence.

  o A **race** is considered simply a subdivision of a species which breeds true except for minor variations.

  o The difference between some breeds is mostly for a trademark and may have little or no economic significance.

Subspecies, Breed, Race and Variety

- A **subspecies** can be considered a natural or wild variation and may be defined by geography.

- A **breed** is synonymous with race (race = breed). Some livestock breeds are:

  o Horses: Arabians, Quarter Horse, Thoroughbred

  o Cattle: Holstein, Brahman, Hereford

  o Pigs: Duroc, Berkshire, Hampshire

  o Sheep: Rambouillet, Suffolk, Hampshire

Chapter Resource

- Goats: LaMancha, Nubian, Alpine

- A variety is a group within a species or subspecies which differs in some significant inherited respect from other members of the species and is usually fertile with any other member of the species.

Scientific Naming of Animals

- **Binomial nomenclature** (also called binary nomenclature) is a formal system of naming species of living things by giving each a name composed of two parts, both of which use Latin grammatical forms.

- Such a name is called a binomial name or a **scientific name**.

- First part of the name identifies the genus to which the species belongs; the second part identifies the species within the genus.

- Since each species has a universal specific name, most individuals know exactly which animal is being referred to when a scientific name is given.

Bos taurus (cow)

Scientific Names for Farm Animals

- Poultry

  - Duck – Domestic - *Anas platyrhyncha*

  - Guinea – *Numida meleagris*

  - Domestic Chicken – *Gallus domesticus*

  - Turkeys – *Melleagris gallopavo*

- Equine

  - Horse – *Equus caballus*

  - Donkey – *Equus asinus*

- Cattle

  - European Breeds – *Bos taurus*

- - Tropical Breeds – *Bos indicus*

    - ✓ A number of cattle breeds which exist that are hybrids of *Bos taurus* and *Bos indicus*. Examples include the Sanga type cattle from Africa which developed naturally hundreds of years ago and the synthetic breeds developed in the 1900's.

- Other Domestic Livestock

  - Sheep – *Ovis aries*
  - Goat – *Capra hircus*
  - Swine – *Sus scrofa*
  - Rabbit – *Oryctolagus cuniculus*

- Other Species

  - American Bison – *Bison bison*
  - Dog – *Canis familiaris*
  - Cat – *Felis catus*

Goat – *Capra hircus*

Summary

Species is a basic group forming a division of genus. Related species which cannot interbreed because of geographical separation are called allopatric species. Those which can interbreed, but in practice do not because of differences in behavior, breeding, food sources, etc., are called sympatric species. Subspecies is a subdivision or smaller part of a group of animals (those in a species). Breed is a race or variety of an animal species, especially one developed through human influence. A subspecies can be considered a natural or wild variation and may be defined by geography. A variety is a subdivision of a species and is usually fertile with any other member of the species. Scientific naming of animals (also called binominal nomenclature or binary nomenclature) is a formal system of naming species of living things by giving each a name composed of two parts, both of which use Latin grammatical forms.

Additional Resources

Animals and Livestock- National Agricultural Library
https://www.nal.usda.gov/animals-and-livestock

Oklahoma State Department of Animal Science
http://www.ansi.okstate.edu/resources/general

Speciation: Of Ligers & Men
https://www.youtube.com/watch?v=2oKlKmrbLoU

Assessment

 Take assessment online here: http://tinyurl.com/AnSci-Species
Download and print the assessment by scanning this QR code or by going to this URL: http://www.tagmydoc.com/Ch03AnSci

4 Animal Protein

Major Concept
Livestock provide protein for the human diet.

Objectives
- Identify major sources of animal protein
- Name the types of livestock operations involved in producing animal protein
- Identify trends in livestock and poultry production

Key Terms
- Aquaculture
- Broilers
- Chevon
- Essential amino acids
- Meat
- Mutton
- Protein
- Veal

Chapter Resource

 Complementary *full color* illustrations, photos, charts and graphs are available by scanning this QR code or by following this URL: http://www.tagmydoc.com/AS04 This digital resource will enhance your understanding of the chapter concepts.

Meat and Protein

- **Meat** is the tissues of the animal body used for food.
  - A nutrient dense food and valuable source of vitamins and mineral.
  - About 15-20% protein
- **Protein** is composed of amino acids necessary for human diets.
- Animal proteins contribute the essential amino acids to the human diet.
  - **Essential amino acids** are those that cannot be made by the body and must be supplied in the diet.

Animals as a Source of Food

- Livestock contributions to the total world protein demand.
    - Livestock products (meat, milk): about 33%
    - Eggs: less than 10%
    - Fish and shellfish: about 33%
- Many developing nations consume large quantities of aquatic animals, many of which are farmed.
- Total amount of protein consumed in developing countries is considerably less than that consumed in Western nations.

Meats

- Red meats include
    - Beef from cattle
    - **Veal** from calves 3 months or younger
    - Pork from swine/pigs/hogs
    - **Mutton** from mature sheep
    - Lamb from young sheep
    - **Chevon** or goat meat

Chapter Resource

- Other meats include:
    - **Broilers** (chickens bred and raised specifically for meat production)
    - Turkeys
    - Game birds
    - Fish and shellfish
- Similar species of animals all around the world.
- In many countries, livestock operations are small and less specialized than those in the U.S.

- Typical meat consumption in the U.S. in lbs/person per year:
    - Beef – 59 to 60 lbs/yr
    - Pork – 47 to 49 lbs/yr
    - Chicken/broiler – 83 to 85 lbs/yr
    - Fish and shellfish – 12 to 16 lbs/yr
        - ✓ Each year, a larger share of the per capital consumption of fish and shellfish in the U.S. comes from aquaculture.

Other Protein Sources

- Typical per capita dairy and egg consumption in the U.S. is:
    - Eggs: 250 per year
    - Dairy products (http://www.ers.usda.gov/data-products/food-availability-(per-capita)-data-system.aspx#.Ur3CMvRDtG4)
        - ✓ Milk and cream: 205 lbs
        - ✓ Butter: 5 lbs
        - ✓ Cheese: 33 lbs
        - ✓ Cottage cheese: 2.3 lbs
        - ✓ Frozen dairy products: 24 lbs
        - ✓ Evaporated and condensed: 7 lbs
        - ✓ Dry milk: 3.5 lbs

Chapter Resource

Basic Animal Science Enterprises to Produce Protein

- Cattle
  - Purebred registered
  - Cow-calf commercial
  - Stocker
  - Feedlot
  - Dairy

- Swine
  - Purebred
  - Breeding
  - Market
  - Specific pathogen-free program
  - Confined areas operations

- Sheep
  - Purebred
  - Breeding
  - Range operations
  - Market
  - Feed lot
  - Small farm operations (farm flock)

- Poultry
  - Broilers/Fryers
  - Layers
  - Turkeys

- Breeding
  - Game birds
- Aquaculture

U.S. Meat Animal Production in Billions of Pounds

- In recent years, the U.S. produced about 26 billion lbs of beef, 49 billion lbs of broilers, 32 billion lbs of hogs and 7-8 billion lbs of turkeys each year.

- Historical perspective of the production of cattle, broilers, hogs and turkey production from the statistics of the USDA – NASS (National Agricultural Statistics Service)

  - http://www.nass.usda.gov/Charts_and_Maps/Meat_Animals_PDI/

- Sheep and lamb inventory declined over the years.

- Not much lamb is eaten in the U.S. – less than one pound per person per year.

- Numbers of Sheep and Lambs from the USDA – NASS.

  - http://www.nass.usda.gov/Charts_and_Maps/Sheep_and_Lambs/sheep1

U.S. Milk Production

- About 9.2 million milk cows in U.S.

- Numbers of milk cows have varied over the years according to USDA – NASS.

  - http://www.nass.usda.gov/Charts_and_Maps/Milk_Production_and_Milk_Cows/milkcows

- Milk production in the U.S. is around 200 billion pounds per year.

- Milk production per cow has increased steadily for the past 10 years according to USDA – NASS.

  - http://www.nass.usda.gov/Charts_and_Maps/Milk_Production_and_Milk_Cows/milkprod

- On average, each cow produces about 21,500 lbs of milk each year.

  - http://www.nass.usda.gov/Charts_and_Maps/Milk_Production_and_Milk_Cows/cowrates

U.S. Poultry and Egg Production

- Broiler production is about 60 billion lbs per year in the U.S.

- USDA - NASS historical data show an increase for a number of years with a leveling off the last few years.

    o http://www.nass.usda.gov/Charts_and_Maps/Poultry/brlprd

- About 250 million turkeys are raised each year.

- According to the USDA - NASS turkey production increased steadily beginning in the late 1960s and early 1970s.

    o http://www.nass.usda.gov/Charts_and_Maps/Poultry/tkyprd

Chapter Resource

- Each year about 93 billion eggs are produced.

- Egg production began increasing around 2008 according to the USDA – NASS.

    o http://www.nass.usda.gov/Charts_and_Maps/Poultry/eggprd

Value of Livestock Production to U.S.

- Meat animal production in the U.S. is valued at about $100 billion with cattle contributing about $50 billion, broilers about $25 billion and hogs about $20 billion.

- Statistics from USDA – NASS show the historical trends for the value of meat animal production.

    o http://www.nass.usda.gov/Charts_and_Maps/Meat_Animals_PDI/valprd

- Total poultry production in the U.S. is valued at about $38 billion and egg production per year is valued at about $8 billion.

- Data from USDA – NASS provides a historical perspective on the value of poultry production.

    o http://www.nass.usda.gov/Charts_and_Maps/Poultry/valprdbetc

Aquaculture

- **Aquaculture** is the production of aquatic animals and plants under controlled conditions for all or parts of their lifecycles.

- Interest is rising because restrictions on the wild harvest of many seafood species may diminish wild harvest seafood supplies.

- During the last two decades, the value of U.S. aquaculture production rose to nearly $1 billion.

- Cultured species include:
  - Catfish
  - Trout
  - Mollusks
  - Shrimp
  - Crayfish

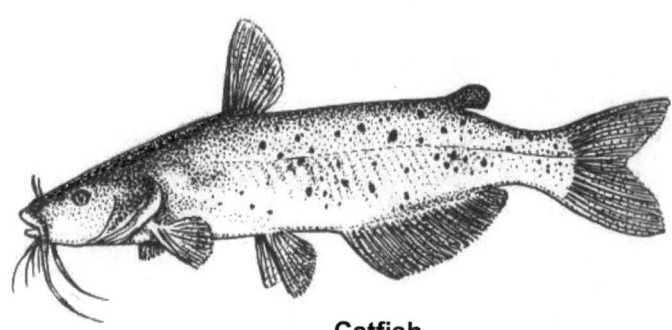

Catfish

- Data on U.S. aquaculture production is tracked by the USDA – ERS.
  - http://www.ers.usda.gov/data-products/aquaculture-data.aspx#.Ur24XfRDtG4

Summary

Meat is the tissues of the animal body used for food. Protein is composed of amino acids necessary for human diets. Animal proteins contribute the essential amino acids to the human diet. Many developing nations consume large quantities of aquatic animals, many of which are farmed. Total amount of protein consumed in developing countries is considerably less than that consumed in Western nations. Red meats include: Beef, veal, pork, mutton, lamb and chevon or goat meat. Other meats include: Broilers, turkeys, game birds and fish and shellfish. Other protein sources include dairy and egg consumption. In recent years, the U.S. produced about 26 billion lbs of beef, 49 billion lbs of broilers, 32 billion lbs of hogs and 7-8 billion lbs of turkeys each year. On average, each cow produces about 21,500 lbs of milk each year. Each year about 93 billion eggs are produced. Meat animal production in the U.S. is valued at about $100 billion with cattle contributing about $50 billion, broilers about $25 billion and hogs about $20 billion. Aquaculture is the production of aquatic animals and plants under controlled conditions for all or parts of their lifecycles. During the last two decades, the value of U.S. aquaculture production rose to nearly $1 billion.

Additional Resources

Taylor, R.E. and T.G. Field. 2010. Scientific farm animal production: Introduction to animal science. 10th ed. Upper Saddle River, NJ: Prentice Hall.

Beef Industry Statistics
http://www.beefusa.org/beefindustrystatistics.aspx

USDA Cattle and Beef Statistics
http://tinyurl.com/kj3s3ur

USDA – National Agricultural Statistics Service (NASS) Charts and Maps
http://www.nass.usda.gov/Charts_and_Maps/index

Assessment

Take assessment online here: http://tinyurl.com/AnSci-AnimalProtein
Download and print the assessment by scanning this QR code or by going to this URL: http://www.tagmydoc.com/Ch04AnSci

5 Meats

Major Concept

Quality and yield grades of livestock are important to producers and consumers.

Objectives

- List the USDA grades for beef, pork and lamb
- Identify wholesale and retail cuts of beef, pork and lamb
- Explain how the USDA criteria are used to grade beef, pork and lamb
- List the dressing percentages of the major commercial meat animals

Key Terms

- Choice
- Commercial
- Cull
- Cutability
- Dressing percentage
- Marbling scores
- Maturity scores
- Prime
- Select
- Standard grade

Chapter Resource

Complementary *full color* illustrations, photos, charts and graphs are available by scanning this QR code or by following this URL: http://www.tagmydoc.com/AS05 This digital resource will enhance your understanding of the chapter concepts.

Carcass Grades

- Beef: More than two types of grades:

- Quality grades, reflect differences in eating quality of meat based on:

    o **Marbling scores** – amount of fat interspersed in the muscle.

    o **Maturity scores** – reflects age of animal at slaughter.

- Types of quality grades for beef are:

    o **Prime** – Superior marbling, proper carcass conformation and adequate maturity.

- ✓ Found in fine restaurants and gourmet stores
- ✓ Beef of this grade is not economical for the meat packer because the cattle are required to get very fat to obtain enough marbling and only a small percentage of cattle meet the conformation standards.

o **Choice** – Most economical and most desirable carcass grade.
- ✓ Adequate marbling and carcass conformation are required.

o **Select** – May be referred to as "no-roll" since it isn't stamped with the USDA grade.
- ✓ Must have slight marbling.
- ✓ Meat is inspected but not marked with a stamp as Prime and Choice, so it can be sold under store names.

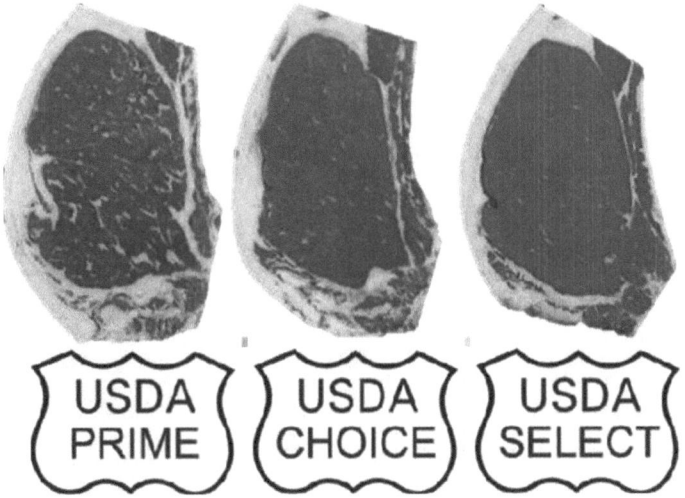

Chapter Resource

o **Standard grade** – Usually older animals and thin animals.
- ✓ Minimum marbling or below average carcass conformation fit into this category.

o **Commercial** – Includes designations of: cutter, canner and utility.
- ✓ Usually processed into lunch meats, soup and canned meat products.

o **Cull** – Not acceptable for human consumption.

- Yield grades

- Identify carcasses for differences in **cutability** (the lean yield of a beef carcass) or yield of boneless, trimmed retail cuts. Yield grade is determined by the following measurements:

  ✓ Hot carcass weight

  ✓ External fat (measured as back fat over the 13th rib)

  ✓ Percent heart, kidney and pelvic fat

  ✓ Rib eye area

Yield Grade	% BCTRC
1	> 52.3
2	52.3 – 50.0
3	50.0 – 47.7
4	47.7 – 45.4
5	< 45.4

- Types of yield grades: 1,2,3,4,5

  ✓ Yield grade 1 being the leanest, heaviest muscled carcass and yield grade 5 being the lightest muscled fattest.

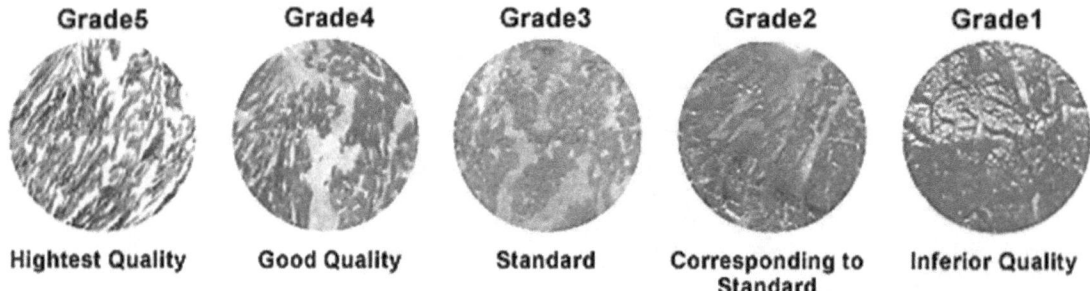

- Sheep

  - Quality Grades – Similar to beef but depend more on maturity

  - Grades for lamb include Prime, Choice, Select and Commercial.

  - Three maturity grades are: 1. Lamb, 2. Yearling mutton, 3. Mutton

  - Consumers in the U.S. demand almost exclusively, lamb.

  - Most lambs are grade Choice.

Chapter Resource

- Swine

  - Quality Grades – two considerations in quality grading swine carcasses are:

    ✓ Quality of lean meat

- ✓ Belly thickness

  o Grades for carcasses that have acceptable lean quality are listed below. The different grades are determined by backfat thickness, carcass length as well as yield of lean cuts. The grades are:

  - ✓ U.S. No. 1
  - ✓ U.S. No. 2
  - ✓ U.S. No. 3
  - ✓ U.S. No. 4

  o Carcasses having unacceptable lean or bellies that are too thin are graded U.S. Utility. Sows also fit this category.

  o Carcasses from boars or stags have a strong "sex" odor and are not passed for use as human food.

Dressing Percent

- **Dressing percentage** is the percentage yield of chilled carcass in relation to the weight of the live animal.

- Beef Cattle

  o Dressing percent depends on the quality grade of the animal.

  - ✓ Prime – 62%
  - ✓ Choice – 60%
  - ✓ Select – 59%
  - ✓ Standard – 57%

  Chapter Resource

  o Example – a 1000 lb prime steer would produce a 620 lb carcass. (The head, feet, hide, internal organs and some fat trim make up the rest of the 380 lbs).

- Swine

  o Dressing percentage of swine is the percentage yield of chilled carcass.

  - ✓ U.S. No.1 – 70%
  - ✓ U.S. No.2 – 71%
  - ✓ U.S. No.3 – 72%
  - ✓ U.S. No.4 – 73%

- ✓ Utility – 69%
  - o Poorer yield grades have higher percentages. This is because the fatter hogs will produce a heavier carcass in relation to their live weight.
- Hogs may be dressed in two ways:
  - o Packer style – with head, kidneys and leaf fat removed.
  - o Shipper style – with head, kidneys and leaf fat left these hogs dress 4 to 8% higher since there is more weight.
- Lamb
  - o Dressing percentage of lamb is the percentage yield of the chilled carcass. Usually lambs are sheared before slaughtering.
    - ✓ Prime – 52%
    - ✓ Choice – 50%
    - ✓ Select – 47%
    - ✓ Utility – 45%

Chapter Resource

Wholesale and Retail Cuts

- Wholesale cuts are larger cuts of meat that are shipped to grocery stores and meat markets.
- For example, a fore quarter or chuck contains roasts, stew meat, etc.
- Retail cuts are family sized or single serving cuts purchased at the market.

Wholesale and Retail Cuts of Beef

Wholesale	Retail
Chuck	ground beef, stew meat, pot-roast, blade roast, short ribs, boneless chuck roast
Rib	rib eye roast or steak
Short loin	top loin steak, T-bone steak, porterhouse steak, filet mignon
Sirloin	sirloin steaks or roast
Round	eye of round, ground beef, cubed steak, rump roast, top round steak, round steak
Fore shank	stew meat, shank cross cuts
Brisket	brisket (corned beef if cured)
Short Plate	short ribs, ground beef, stew meat
Flank	ground beef, flank steak

Wholesale and Retail Cuts of Pork

Wholesale	Retail
Jowl	smoked jowl
Picnic shoulder	fresh arm roast, sausage, neck bones, fresh or smoked picnic shoulder
Belly	spare ribs, salt pork, bacon
Leg	fresh leg or ham
Loin	center loin roast, sirloin roast, tenderloin, back ribs, blade chops, loin chops, Canadian style bacon
Boston butt	blade roast, blade steak

Wholesale and Retail Cuts of Lamb

Wholesale	Retail
Shoulder	neck slices, arm chops, blade chops, stew meat, shoulder roast
Rib	rib roast, rib chops
Loin	loin roast, loin chops, sirloin chops
Leg	leg (many styles of preparing - French style, American style, boneless)
Breast	riblets, breast
Shank	fore and hind shanks

Summary

Beef: Two types of grades, quality grades, reflect the differences in the eating quality of meat based on: marbling scores and maturity scores. Types of quality grades: Prime, choice, select, standard, commercial and cull. Yield grades identify carcasses for differences in "cutability" or yield of boneless, trimmed retail cuts. Yield grade 1 being the leanest, heaviest muscled carcass and yield grade 5 being the lightest muscled and fattest. Sheep quality grades are like beef but depend more on maturity. The grades for lamb include prime, choice, select and commercial. Swine have two considerations in quality grading swine carcasses: quality of the lean meat and belly thickness. The dressing percentage is the percentage yield of chilled carcass in relation to the weight of the live animal.

Additional Resources

Aberle, E.D., J.C. Forrest, D.E. Gerrard and E.W. Mills. 2013. Principles of meat science. 5th ed. Dubuque, IA: Kendall/Hunt Publishing.

Beef Grading-Swine Grading and Sheep Grading links:
http://www.sdstate.edu/ars/students/activities/judging/evaluation/beef-grading

Grading, Certification and Verification
http://www.ams.usda.gov/AMSv1.0/LSMeatGradingGradingServices

Slaughter Steers Poster
http://www.ams.usda.gov/AMSv1.0/getfile?dDocName=STELDEV3067087

What's Your Beef-Prime, Choice or Select?
http://tinyurl.com/nznwok6

Assessment

Take assessment online here: http://tinyurl.com/AnSci-Meats
Download and print the assessment by scanning this QR code or by going to this URL: http://www.tagmydoc.com/Ch05AnSci

6 Hormones and Meat Production

Major Concept

Hormones and hormone-like substances used as growth promotants in the livestock industry can create public health issues.

Objectives

- Define steroid, half-life and growth promotants
- Identify hormones used in meat animal production
- List four growth promotants and identify how they work in the animal body

Key Terms

- Diethylstilbestrol (DES)
- Estradiol benzoate
- Estrogen
- Growth promotants
- Half-life
- Hormone
- Melengestrol acetate (MGA)
- Norgestomet
- Progesterone
- Steroid
- Synovex H
- Synthetic
- Testosterone propionate
- Zeranol (Ralgro)

Chapter Resource

 Complementary *full color* illustrations, photos, charts and graphs are available by scanning this QR code or by following this URL: http://www.tagmydoc.com/AS06 This digital resource will enhance your understanding of the chapter concepts.

What is a Hormone?

- A chemical substance produced by the endocrine glands.

  o Released into the bloodstream and transported to target cells where it regulates some bodily function or influences an activity or behavior.

  o Target cells are cells which recognize only a particular hormone or hormones.

What is a Steroid?

- A lipid (fat) manufactured from cholesterol in the body.

  o Sex hormones (testosterone, estrogen and progesterone) are steroids.

    ✓ **Estrogen**: Causes estrus and the development of secondary female sexual characteristics.
    ✓ **Progesterone**: The pregnancy hormone helps prepare the female's body for and then maintain pregnancy.
    ✓ **Testosterone**: Is produced in the testes and causes the development of secondary male sexual characteristics.

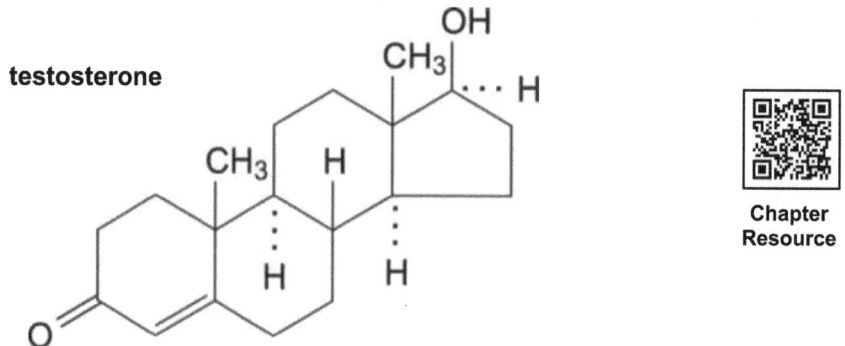

Chapter Resource

- Steroids may be natural (made by the body) or **synthetic** (made in a laboratory).

  o Synthetic hormones are stronger than natural ones.

- Synthetic steroids are the type used in the livestock industry as **growth promotants** (used to help increase the efficiency of animal production by increasing weight gain and product output).

Synthetic Steroids Used in Meat Production

- **Zeranol (Ralgro)** – A synthetic compound (not a steroid) and is recognized in the target cells as estrogen.

  o Implanted behind the ear, where it is absorbed through the skin.

  o Promotes nitrogen retention, which encourages protein synthesis and muscle mass, which therefore increases growth rate.

  o One of the most widely used growth promoters.

  o A similar substance is **Synovex H** (for heifers) and **S** (for steers).

- ✓ In a study, use of Synovex H showed a 17 lb weight increase in heifers and a 9-14 lb increase in steers of slaughter weight over those not treated with Synovex H/S.

- **Estradiol Benzoate** (artificial estrogen) – A natural estradiol (type of estrogen) combined with the chemical benzoate.

  - Purpose of this combination is to slow down the absorption of the steroid from the injection site for a "time-release" effect. This allows for fewer injections.

- **Melengestrol Acetate** (MGA) – A growth promotant usually given as an implant, although it can also be given orally.

  - Effective only in females.

    - ✓ A progesterone-like compound (a female steroid) which allows the female to grow her ovarian follicles but inhibits ovulation. Thus, the female produces her own estrogen as a growth promotant.

    - ✓ Once used to synchronize heat because it suppresses estrus as long as it is used. Once it is stopped, the heifer comes into heat.

  - Approved in 1967 for oral use in cattle rations.

- **Diethylstilbestrol (DES)** – A synthetic (estrogen) compound recognized by the estrogen receptors as a steroid.

  - Promotes growth in immature animals.

    - ✓ Increased growth is in lean (muscle) content, not fat.

  - Approved in 1954 for use in cattle feed.

Chapter Resource

  - Used as a growth promotant but has since been removed from the market.

- **Testosterone Propionate** – A natural hormone which has been combined with a chemical, propionic acid, to increase its **half-life** (the time it takes for the body to eliminate half of the substance - it is a common measure for use in describing how long substances stay in an animal's body).

  - Sometimes combined with estradiol benzoate in implants to promote more rapid growth.

- **Norgestomet** – Same effect as MGA

Misuse of Hormones

- Improper use of any chemical substance is a hazard to humans.

- Body produces some of the same steroids as the synthetic ones used as growth promotants.

- If the steroid treatment given to meat animals is mismanaged the combination of natural human steroids and synthetic ones from the meat can cause an excess in the human body, which becomes a health hazard.

    - Two of the most serious hazards for humans include higher incidence of cancer or birth defects.

Summary

Hormones and hormone-like substances are used as grow promotants in the livestock industry. A hormone is a chemical substance produced by the endocrine glands. A steroid is a lipid (fat) manufactured from cholesterol in the body. The sex hormones (testosterone, estrogen and progesterone) are steroids. Zeranol is a synthetic compound and is recognized in the target cells as estrogen. Estradiol Benzoateis natural estradiol (type of estrogen) combined with the chemical benzoate. The purpose of this combination is to slow down the absorption of the steroid from the injection site for a "time-release" effect. MGA is also a growth promotant. DES is a synthetic (estrogen) promotes growth in immature animals but is no longer used. Testosterone Propionate is a natural hormone which has been combined with a chemical, propionic acid, to increase its half-life (the time it takes for the body to eliminate half of the substance. Improper use of any chemical substance is a hazard to humans.

Additional Resources

Field, T.G. and R.E. Taylor. 2012. Scientific farm animal production: An introduction to animal science. Upper Saddle River, NJ: Prentice Hall.

EPA, Ag101, Beef production
http://www.epa.gov/oecaagct/ag101/printbeef.html

Stewart, L. 2010. Implanting beef cattle
http://extension.uga.edu/publications/detail.cfm?number=B1302#1

Assessment

 Take assessment online here: http://tinyurl.com/AnSci-Hormones1 or http://tinyurl.com/AnSci-Hormones-MeatProduction
Download and print the assessment by scanning this QR code or by going to this URL: http://www.tagmydoc.com/Ch06AnSci

7 Milk and Eggs

Major Concept

Milk and eggs, products of livestock production, provide important sources of nutrition for human diets.

Objectives

- Identify the steps in milk production
- Define the following terms: pasteurized milk ordinance, candling, somatic cell count
- Compare the nutritive value of milk and eggs to other foods
- Identify the classes of milk and their uses
- Recognize the scientific principles involved in the processes of pasteurization and homogenization
- Name three methods of pasteurization
- Outline the processing functions of the egg industry
- List steps in the sizing, grading and packaging of eggs
- Identify the steps in the proper care and handling of eggs
- Identify the components needed for the proper labeling of eggs

Key Terms

- Candling
- Casein
- Colloidal suspension
- Contaminants
- Designer eggs
- Homogenized
- Julian date
- Lactalbumin
- Lactoglobulin
- Lactose
- Lipids
- Pasteurized
- Pasteurized Milk Ordinance
- Plant code
- Somatic cell count

Chapter Resource

Complementary *full color* illustrations, photos, charts and graphs are available by scanning this QR code or by following this URL: https://www.tagmydoc.com/AS07Combined This digital resource will enhance your understanding of the chapter concepts.

Milk Production

- Dairy cows produce over 90% of consumed milk in the US.
- Milk from buffalo and goats is increasing.
- Milk from sheep is decreasing.

Major Sources of World Milk Production

- Cows
- Buffalo
- Goats
- Sheep

Chapter Resource

Milk Production Steps

- Milking

  o Obtained from the cow under sanitary conditions and cooled to 45°F (7°C) within 2 hours of milking.

  o Producers of milk pay for the volume shipped, and for the nutrient value of that milk (% fat, % protein or % solids-not-fat) and for the "quality" of the milk as indicated by **somatic cell count** (an indicator of the quality of milk) higher count lower quality.

- Stored in farm tanks

- Trucked and sampled

  o Milk is picked up by a handler who takes a sample and then pumps the milk from farm's bulk tank into the milk truck.

  o Before the milk can be unloaded at the processing plant, each load is tested for antibiotic residues.

  o If the milk shows no evidence of antibiotics, it is pumped into the plant's holding tanks for further processing.

- If the milk does not pass antibiotic testing, the entire truck load of milk is discarded and the farm samples are tested to find the source of the antibiotic residues.

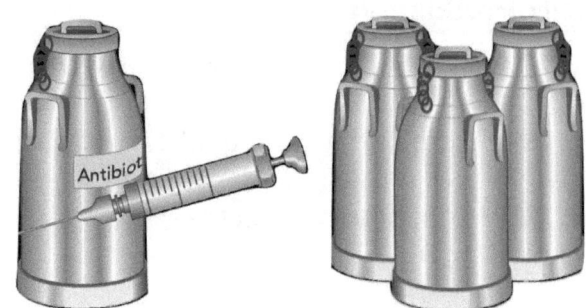

- Regulatory action is taken against farms with a positive antibiotic test.

- Positive antibiotic tests are rare and account for less than 1% of the tank loads of milk delivered to processing plants.

- Processed

    - Milk at the plant is stored at less than 45°F (7°C) and is usually processed within 24 hours, but can held for up to 72 hours (3 days) before processing.

    - **Pasteurized** – milk that has been exposed briefly to high temperatures to destroy microorganisms and prevent fermentation (process described below).

    - **Homogenized** – milk with the fat particles broken up and dispersed uniformly so the cream will not rise (process described below).

    - Variety of products, including fluid milk, cheese, ice cream and yogurt.

    - Composition of many dairy products is defined by law, called Standards of Identity, in the United States Code of Federal Regulations (2006).

- Packaged

- Delivered to stores

Milk Grade

- **Pasteurized Milk Ordinance** (PMO, 2005) provides standardized guidelines.

    - PMO is a document from the United States Departments of Health and Human Services and Public Health and the Food and Drug Administration that defines practices relating to milk parlor and processing plant design, milking practices, milk handling, sanitation and standards for the pasteurization of Grade A milk products.

    - Each state regulates their own dairy industry, but the state's guidelines usually meet or exceed those defined by the PMO.

- Non-Grade A milk would either be considered as manufacturing grade (sometimes called Grade B) or rejected.

Nutritive Value of Milk

- Termed a **colloidal suspension** (a colloid that has a continuous liquid phase in which a solid is suspended in a liquid.)

    o   Whole milk is:

    Chapter Resource

    ✓ 88% water

    ✓ 8.6% solids not fat

    ✓ 3-4% fat (sometimes called butter fat).

- High nutrient density

- Concentration of important nutrients high in relation to relatively low calories

- Contains nutrients such as fat, protein, carbohydrates, vitamins and minerals

Milk Fat

- Mixture of **lipids** (any one of various substances that contain fat) existing as microscopic globules suspended in the milk.

- Contributes about 48% of the calories in whole milk.

- Contains the fat-soluble vitamins (A, D, E and K).

- Contains the flavor components of milk.

- Made of short and long chain fatty acids.

Carbohydrates

- Predominate is **lactose** (milk sugar).

- Contributes about 30% of the total calories in milk.

- About 4.8% of cow's milk is lactose.

Proteins

- Approximately 3.3% of milk is protein.

- Accounts for about 22% of the calories of whole milk.

- **Casein**, a protein only found in milk, is about 82% of the total milk protein.

- Whey proteins, primarily **lactalbumin** (simple protein of milk) and **lactoglobulin** (a crystalline protein fraction) make up remaining 18% of milk protein.

Vitamins

- Contains all vitamins (fat-soluble and water-soluble) essential to human nutrition.

- Fat-soluble: A, D, E and K (depends on fat content).

- Water-soluble: not a good source of C or all the B vitamins.

Chapter Resource

Minerals

- Rich source of calcium

- Good source of phosphorus and zinc

- Not a good source of iodine

Homogenization

- Fat in milk secreted by the cow in globules of non-uniform size, ranging from 0.20 to 2.0 µm (micrometer or one-millionth of a meter).

- Non-uniform size of globules causes them to float, or cream, to the top of the container.

- Pasteurized milk does not necessarily need to be homogenized but homogenized milk should be pasteurized to inactivate native enzymes that deteriorate fat (lipases) and cause rancidity, which results in off-flavors and reduced shelf life in milk.

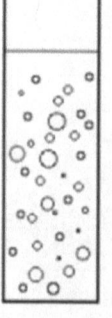
Raw milk

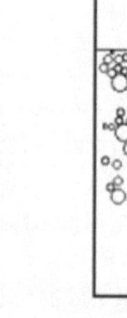
Cold, raw milk after 1 hour

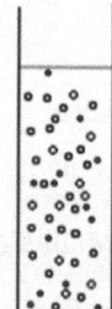
Homogenized milk during storage

- Purpose of homogenization to reduce the milk fat globules size to less than 1.0 μm allowing them to stay evenly distributed in milk.

- High pressure process that forces milk at a high velocity through a small orifice to break up the globules, creating many more fat globules of a smaller size.

Pasteurization

- Process of heating a liquid to below the boiling point to destroy microorganisms depends on temperature and time held at temperature.

- Most all the milk in the United States is pasteurized, except for some raw milk cheese production and some states that allow the sale of raw milk.

- Conditions of the heat treatment used for pasteurization depend on the final product – lower temperatures are used for refrigerated products and higher heat treatments are used for products stored at room temperature.

- Table of the types/methods of pasteurization:

Pasteurization Type	Typical Product	Typical Storage	Temperature	Holding Time
Batch, vat	Milk	Refrigerated	145°F (62.8°C)	30 min
Batch, vat	Viscous products, or products with more than 10% fat or added sweetener	Refrigerated	150°F (65.6°C)	30 min
Batch, vat	Egg nog, frozen dessert mixes	Refrigerated	155°F (68.3°C)	30 min
Continuous, high temperature short time (HTST)	Milk	Refrigerated	161°F (71.7°C)	15 sec
Continuous, high temperature short time (HTST)	Viscous products, or products with more than 10% fat or added sweetener	Refrigerated	166°F (74.4°C)	15 sec
Continuous, high temperature short time (HTST)	Egg nog, frozen dessert mixes	Refrigerated	175°F (79.4°C)	25 sec
Continuous, high temperature short time (HTST)	Egg nog, frozen dessert mixes	Refrigerated	180°F (82.2°C)	15 sec
Continuous, higher heat shorter time (HHST)	Milk	Refrigerated	191°F (88.3°C)	1 sec
Continuous, higher heat shorter time (HHST)	Milk	Refrigerated	194°F (90°C)	0.5 sec
Continuous, higher heat shorter time (HHST)	Milk	Refrigerated	201°F (93.8°C)	0.1 sec
Continuous, higher heat shorter time (HHST)	Milk	Refrigerated	204°F (96.2°C)	0.05 sec
Continuous, higher heat shorter time (HHST)	Milk	Refrigerated	212°F (100°C)	0.01 sec
Continuous, Ultra pasteurization	Milk and cream	Refrigerated, Extended storage	280°F (137.8°C)	2 secs
Aseptic, ultra high temperature (UHT)	Milk	Room temperature	275-302°F (135-150°C)	4-15 secs
Sterilization	Canned products	Room temperature	240°F (115.6°C)	20 min

- Process of heating or boiling milk for health benefits recognized since the early 1800s and used to reduce milkborne illness and mortality in infants in the late 1800s.

- Common milkborne illnesses during turn of the 20th century were typhoid fever, scarlet fever, septic sore throat, diphtheria and diarrheal diseases.

- Virtually eliminated with the commercial implementation of pasteurization, in combination with improved management practices on dairy farms.

Producing Fluid Milk

- To achieve standardization, milk is processed through centrifugal separators to create a skim portion and a cream portion of the milk.

- Separation produces a skim portion that is less than 0.01% fat and a cream portion that is usually 40% fat.

- Cream portion is then added back to the skim portion to yield the desired fat content for the product. Four common categories of fluid milk regarding fat content:

    1. Skim or nonfat (<0.1% fat)
    2. 1% fat
    3. 2% fat
    4. Whole (3.25% fat)

Chapter Resource

Classified Pricing

- Pricing system dictates prices that differ according to category of use.

- In Federal and some state milk marketing orders, regulated processors must pay a minimum price for Grade A milk according to the class in which it is used.

- Most of the states that have their own pricing regulations, including California, have price structures analogous to the current Federal milk marketing orders.

- Four classes (uses):

    o Class I. Grade A milk used in all beverage milks.

    o Class II. Grade A milk used in fluid cream products, yogurts, or perishable manufactured products (ice cream, cottage cheese and others).

- Class III. Grade A milk used to produce cream cheese and hard manufactured cheese.
- Class IV. Grade A milk used to produce butter and any milk in dried form.

Eggs

- Each year about 93 billion eggs are produced. That is 93 with nine zeros or 93,000,000 000! These eggs provide a source of nutrition for humans and they are used in a variety of food products. Commercial egg production involves science, management skills and business skills.

Nutritive Value of Eggs

- One whole, large egg contains the following nutrients:

  - Calories: 78
  - Total fat: 5 g
  - Cholesterol: 187 mg
  - Total carbohydrate: 0.6 g
  - Protein: 6 g (with all essential amino acids)
  - Vitamins: A, D, B$_6$, B$_{12}$
  - Minerals: sodium, potassium, calcium, iron and magnesium

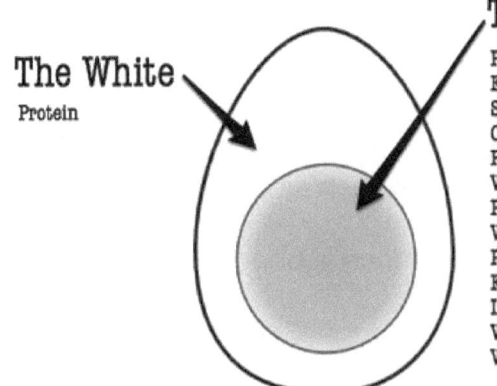

The White
Protein

The Yolk
Protein
Essential Fatty Acids
Selenium
Choline
Riboflavin
Vitamin D
Phosphorus
Vitamin B12
Pantothenic Acid
Folate
Iron
Vitamin A
Vitamin B6

Steps in the Processing of Eggs

- Cleaning the eggs

  - All eggs in the United States are washed in temperature controlled water with a mild detergent and pH adjustments to remove any **contaminants** (presence of a minor and unwanted constituent in a material such as manure, grease, blood, yolk, etc.) before they are sold for human consumption.
  - Eggs are dried to remove excess moisture prior to packaging.
  - Removal of contaminants prevents egg spoilage by bacteria.

- Eggs not clean or containing defects are removed from the processing line and are not packaged for shell egg sale; sent to further processing, including pasteurization.

Grade AA	Grade A	Grade B
Egg content covers a small area. White is firm, has much thick white surrounding the yolk and a small amount of thin white. The yolk is round and upstanding.	Egg content covers a moderate area. White is reasonably firm and has a considerable amount of thick white and a medium amount of thin white. The yolk is round and upstanding.	Egg content covers a very wide area. White is weak and watery, has no thick white and the large amount of thin white is thinly spread. The yolk is enlarged and flattened.

- Grading

  - Grading refers to the process of grouping eggs according to similar characteristics such as quality and weight.

  - Quality assurance is performed by company employees.

  - USDA has a stringent set of requirements for the grading of fresh shell eggs.

  Chapter Resource

  - Egg grading is dependent upon examination of internal quality factors (e.g., condition of the egg white and yolk, air cell size).

    ✓ Internal quality factors are determined by candling.

    ✓ **Candling** involves holding the egg to a concentrated light source for visual inspection of internal defects, such as blood spots, double yolks or air cell size.

  - Egg grading is also dependent upon external quality factors (e.g., shape, texture, cleanliness and soundness of the shell).

  - External quality can be determined by candling for illumination and detection of egg shell cracks.

- An AA grade egg contains the most desirable characteristics while an egg with a B grade contains the least desirable characteristics. The USDA has specifications for egg shell, air cell, egg white and egg yolk characteristics.

Size Determination

- Sizes are determined by weight.

- Six different weight categories: peewee, small, medium, large, extra-large and jumbo. Each size category receives a different price on the farm as well as at the retail level.

    o Minimum average weight for one dozen eggs:

    1. Peewee: 15 oz
    2. Small: 18 oz
    3. Medium: 21 oz
    4. Large: 24 oz
    5. Extra Large: 27 oz
    6. Jumbo: 30 oz

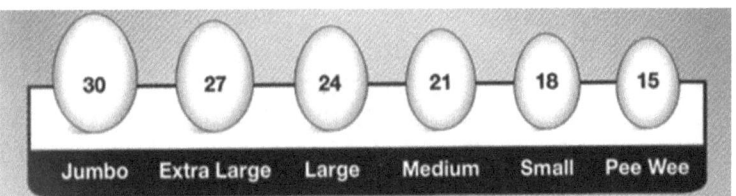

Packaging of Eggs

- Eggs are packaged into all sorts of containers designed for safe shipment of and product appearance for the consumer.

- Loose packaging is a simple way of moving large quantities of eggs in an efficient but also plain fashion.

    o Usually refers to either a thirty-egg large or extra-large flat or a twenty-egg jumbo flat.

    o Loose sales mostly consist of sales to restaurants or other food service consumers.

- In retail settings, eggs are often packaged in cartons containing either one dozen or eighteen eggs. (This packaging method is designed to be handled by the final consumer and is therefore usually designed to be quite appealing).

- Several different varieties of eggs are available to the consumer, which are called specialty or **designer eggs** in which the content has been modified from the standard egg by the industry.

Care and Handling of Eggs

- Mechanical Handling

    o Eggs are moved using conveyor systems between the production facility and the processing plant. Egg flow is normally managed with computerized flow control systems.

    o Processing machinery is fully mechanized and most of the equipment on commercial farms wash, dry, sort by weight and quality and package the eggs into specified packaging.

    o Once eggs are consolidated to pallets, fork lifts are used to handle the product.

Labeling

- Labels must provide certain information: sell by date, Julian date, plant code, grade, nutritional facts, "Keep Refrigerated" and animal care certified.

- Sell by Date

    o Every individual package of eggs processed must contain a sell by date that is set from the date of processing.

- **Julian Date**

    o Three-digit day number relative to the day in the year the eggs were processed. This date is calculated with January 1 as 001 and December 31 as 365.

- **Plant Code**

    o A code is printed on every carton produced in a processing plant. Each plant has its own individual code so that if there is a need to find the origin of the finished product, it can be traced all the way back to the processing plant.

- Grade

    o Both egg size and level of quality are printed on carton, quality is signified by an A or AA grade on almost all fresh shell eggs.

- Nutritional Facts

  o Nutritional information is printed on every package for retail sale. This information varies slightly depending on the egg size and the type of egg (designer eggs).

- "Keep Refrigerated"

  o All egg packages are labeled with a cautionary "keep refrigerated" label. The reasoning for this warning pertains to the effort to reduce growth of salmonella if present.

  o Refrigeration

    ✓ Eggs must be cooled to an internal temperature of 45°F.

    ✓ All graded eggs being transported must be hauled in a refrigerated trailer so that the core temperature is held above freezing but below 45°F.

- U.E.P. "Animal Care Certified"

  o Nationwide, >80% of egg farmers have voluntarily joined the United Egg Producers (UEP), which sets a score of regulations to ensure the overall welfare of chickens being used for egg production. When a farm becomes "Animal Care Certified", this label can be used to inform the consumer that the product they are purchasing was produced in accordance with accepted animal welfare standards.

Summary

Dairy cows produce over 90% of the milk consumed. Major sources of world milk production: Cows, buffalo, goats and sheep. Milk production steps: Milking, stored in farm tank, trucked and sampled, processed, packaged and delivered to stores. Pasteurized Milk Ordinance (PMO, 2005) provides standardized guidelines. PMO is a document from the United States Departments of Health and Human Services and Public Health, and the Food and Drug Administration that defines practices relating to milk parlor and processing plant design, milking practices, milk handling, sanitation, and standards for the pasteurization of Grade A milk products. Whole milk is 88% water, 8.6% solids not fat and 3-4% fat (sometimes called butter fat). Milk fat is a mixture of lipids (of the fats) existing as microscopic globules suspended in the milk and contributes about 48% of the calories in whole milk. Predominate carbohydrate is lactose (milk sugar) and contributes about 30% of the total calories in milk. Approximately 3.3% of milk is protein and accounts for about 22% of the calories of whole milk. Milk contains all vitamins (fat-soluble and water-soluble) essential to human nutrition. Milk is a rich source of calcium and a good source of phosphorus and zinc. Purpose of homogenization is to reduce the milk fat globules size to less than 1.0 μm allowing them to stay evenly distributed in milk. Pasteurization is the process of heating a liquid to below the boiling point to destroy microorganisms, depends on temperature and time held at temperature. To achieve standardization, milk is processed through centrifugal separators to create a skim portion and a cream portion of the milk.

All eggs in the United States are washed in temperature controlled water with a mild detergent and pH adjustments in order to remove any contaminants before they are sold for human consumption. Grading refers to the process of grouping eggs according to similar characteristics, such as quality and weight. Egg grading is also dependent upon external quality factors (e.g., shape, texture, cleanliness and soundness of the shell). External quality can be determined by candling for illumination and detection of egg shell cracks. Egg sizes are determined by weight. Eggs are packaged into all sorts of containers designed for safe shipment of and product appearance for the consumer. Egg flow is normally managed with computerized flow control systems. Eggs are labeled with: Sell By Date, Julian Date, Plant Code, Grade, U.E.P. "Animal Care Certified," "Keep Refrigerated," Refrigeration and Nutritional Facts.

Additional Resources

Taylor, R.E. and T.G. Field. 2011. Scientific farm animal production. 10th ed. Upper Saddle River, NJ: Prentice Hall.

Commercial Egg Production and Processing
http://ag.ansc.purdue.edu/poultry/publication/commegg/

History of Cow's Milk from the Ancient World to the Present
http://milk.procon.org/view.timeline.php?timelineID=000018

Incredible Egg
http://www.incredibleegg.org/

Milk Production
http://usda.mannlib.cornell.edu/MannUsda/viewDocumentInfo.do?documentID=1103

Assessment

 Take assessment online here: http://tinyurl.com/AnSci-MilkandEggs1
Download and print the assessment by scanning this QR code or by going to this URL: http://www.tagmydoc.com/Ch07AnSci

8 Wool

Major Concept

Sheep produce wool and wool quality is influenced by breeds and management.

Objectives

- Define the difference between wool and hair of mammals
- Name one desirable and one undesirable quality of wool
- List the three ways wool can be classed and graded
- Name the two major factors that affect the quality of wool
- Indicate why the National Wool Act was passed

Key Terms

- Cortex
- Crimp
- Cuticle
- Felting
- Follicle
- Kemp
- Medulla
- Wool staple
- Woolen
- Worsted wool

Chapter Resource

Complementary full color illustrations, photos, charts and graphs are available by scanning this QR code or by following this URL: http://www.tagmydoc.com/AS08 This digital resource will enhance your understanding of the chapter concepts.

Value of Hair/Wool from Mammals

- Hair from mammals has limited commercial value for clothing because of the availability of cotton and synthetic products.

- Wool is expensive to produce and process.

- Wool has been used extensively because of its ability to keep heat out or absorb heat.

    o Also, has great moisture absorption ability.

- Hair is used for secondary purposes such as padding.

- Mohair is used in clothing and home décor and as a substitute for fur.

    o Also, blended with other fabrics because of its luster and ability to absorb dyes.

Growth of Wool, Hair and Mohair

- Wool or hair fiber growth takes place in the root bulb of a **follicle** which occurs in the outer layers of the skin.

    o A cluster of follicles creates a cluster of wool fibers.

    o A cluster of wool fibers is called a **wool staple** and the length can determine the use of the wool.

- Growth occurs at the base of the follicle where blood supply is located. The cells produced are pushed outward and die after being removed from the blood supply.

- The fibers have a cuticle and a cortex. The outer layer of the fiber is called a **cuticle** and causes the fibers to cling together. The **cortex** surrounds an inner core or **medulla**. Only medium and coarse wools have a medulla.

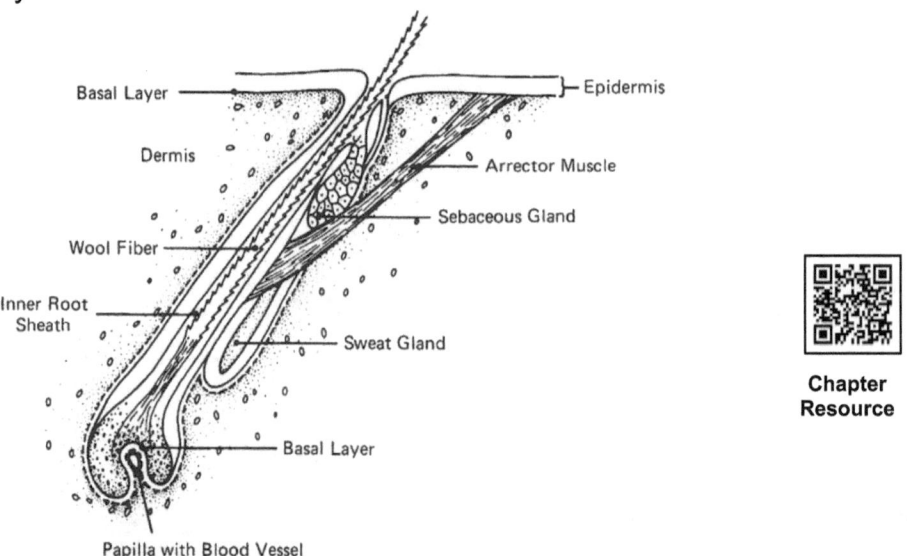

- Intermingling of wool fibers is called **felting**. Felting can make woolens but also shrinks when wet.

- Wool fibers have waves called **crimp**. Some wool fibers called **kemp** have no crimp, are large, black and reduce the value of the fleece. These fibers do not dye well and show in clothing.

Factors Affecting the Value of Wool

- Two major factors that affect the value of wool are nutrition and breeding. The value of the wool can also be improved by how it is handled.

    o Amount of feed.

    o Protein content of feed at 8% or above.

    o Selection of rams for quantity and quality of wool production.

    o Avoid stressing of animals.

    o Example, no food or water or too much heat.

    o Handling of fleece when shorn and prepared for market:

    ✓ Shorn dry

    ✓ Matted, contaminated, inferior portions of wool removed.

    ✓ Keeping different parts of fleece separate such as the neck, jowls, trimmings and back.

    ✓ Product sacked differently for different grades (black-faced, black fibers, dead animals, coarse fibers such as kemp).

    ✓ Use only lanolin-based paint for marking animals.

Chapter Resource

National Wool Act and Incentive Payment

- A permanent price support program was created in 1954 for wool and mohair to increase domestic production.

- Incentive payments – Direct payments made to producers of wool and mohair. This is the percentage needed to bring the national average return to producers up to the annually set national support price; discontinued in 1995.

Classes and Grades of Wool

- Value of wool is determined by class (fineness of fibers) and grade (length of fibers).

- Sheep breeds produce different grades of wool, the Merino breed producing the finest wool.

- o White wool is the preferred color.

- Fiber diameter is the most significant feature which determines the physical characteristics of a finished product.

- More uniform the individual fibers, the more valuable the wool.

- Fineness of the fibers will determine how the wool will be used.

  - o American (blood) grade – relates to the percentage of Merino blood carried by the sheep such as fine, 1/2-blood, 3/8-blood, etc.

  - o Spinning count – based on the number of hanks of yarn which could be spun from one pound of clean wool; provides a numerical designation of fineness such as 80s, 70s, down to 36s; a "hank" of yarn is 560 yards in length.

  - o Micro diameter – Individual fibers are accurately measured which is one micron or one millionth of a meter; fineness being expressed as the average fiber diameter.

Chapter Resource

Uses of Wool and Mohair

- Wool can be processed into other refined products such as felt, tweed, worsted and yarn.

  - o **Woolen** is a type of yarn that is created from carded wool. It is light, soft and stretchy. It can be used to make blankets, hosiery and flannels.

  - o **Worsted wool** is made from the long fibers that have been combed to make sure the fibers run the same direction and not carded but washed.

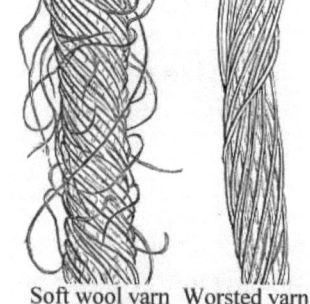

Soft wool yarn   Worsted yarn

    ✓ Coarser and weaved into usually a twill or plain fabric such as gabardine and used for gloves, clothing and pool table coverings.

- Wool in general is used for clothing, horse blankets, saddle cloths, carpets, insulation, and upholstery, heavy insulation covers for odors and noise, baby sleep products, blended with Kevlar for body armor.

- Mohair – scarves, winter hats, suits, sweater, coats, socks home furnishings, carpet yarns and doll wigs.

- U.S. military purchases wool for use in combat clothing, uniforms, dress shirts and pants, socks, underwear, berets, gloves, blankets and fire-resistant fabrics.

Qualities of Wool

- Desirable qualities
  - Warming ability
  - Absorbs considerable amount of moisture.
  - Resistant to fire
- Undesirable qualities
  - Shrinks when wet
  - Can cause itching

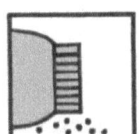
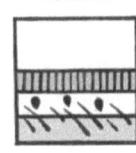

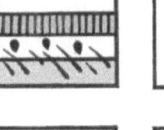

- Continual research on creating washable wool clothing for easier upkeep.

Wool Production in the United States

- Wool is grown in all 50 states with Texas, Wyoming, Montana and California producing the top amounts.

- U.S. imports wool from New Zealand and Australia.

- China buys approximate 70% of US wool exports.

Chapter Resource

- Wool prices have varied widely over the last decade.

- The most up-to-date information can be found on this link: https://www.sheepusa.org/ResearchEducation_MarketReports_UsdaWeekly

- Many smaller-scale farmers produce wool for specialty hand spinning markets.

- Most wool is marketed with private or cooperative wool warehouses. Wool pools are a place for locally marketed wools.

- Most woolen mills are in the eastern United States and create textiles from lightweight worsteds to fancy woolens.

Summary

Hair from mammals has limited commercial value for clothing because of the availability of cotton and synthetic products. Wool is expensive to produce and process. Wool fibers possess unique characteristics. Wool or hair fiber growth takes place in the root bulb of a follicle which occurs in the outer layers of the skin. The two major factors that affect the value of wool are nutrition and breeding. The value of wool is determined by class (fineness of fibers) and grade (length of fibers). Fiber diameter is the most significant feature which determines the physical characteristics of a finished product. The more uniform the individual fibers, the more valuable the wool. The fineness of the fibers will determine how the wool will be used. Wool can be processed into other refined products such as felt, tweed, worsted and yarn. Wool in general is used for clothing and a variety of other products. Desirable qualities of wool include: warming ability, absorbs considerable amount of moisture, resistant to fire. Undesirable qualities include: shrinks when wet and can cause itching. Wool is grown in all 50 states with Texas, Wyoming, Montana and California leading in production.

Additional Resources

Agricultural Marketing Resource Center - Wool
http://www.agmrc.org/commodities__products/livestock/lamb/wool/

American Sheep Industry Association - American Wool Council
http://www.sheepusa.org/

American Sheep Industry Association – Wool Information
http://www.sheepusa.org/Wool_Information

Angora
http://www.ansi.okstate.edu/breeds/goats/angora/index.htm

Colorado State University Extension - Grades and Lengths of Grease Wool
http://www.ext.colostate.edu/pubs/livestk/01401.html

Real Men Wear Wool
http://www.sheep101.info/wool.html

Assessment

 Take assessment online here: http://tinyurl.com/AnSci-Wool
Download and print the assessment by scanning this QR code or by going to this URL: http://www.tagmydoc.com/Ch08AnSci

9 Animal Cells

Major Concept
Animal cells are the basic units of life.

Objectives
- List organelles of an animal cell and identify their function
- Identify three main functions of most cells
- Name four types of cells

Key Terms
- ADP
- ATP
- Cell wall
- Cellulose
- Centriole
- Centrosome
- Differentiate
- Endoplasmic reticulum
- Eukaryotes
- Golgi apparatus/bodies
- Lysosomes
- Mitochondria
- Nucleus
- Organelle
- Prokaryotes
- Ribosomes
- Vacuole

Chapter Resource

Complementary *full color* illustrations, photos, charts and graphs are available by scanning this QR code or by following this URL: http://www.tagmydoc.com/AS09 This digital resource will enhance your understanding of the chapter concepts.

Three General Functions of Most Cells

1. Maintenance

2. Synthesis of cell products

3. Cell division

General Cell Structure

- Cell is the basic structure of life within which chemical reactions of life occurs providing metabolites for human use.

- **Prokaryotes** – cells with no true nucleus.

- **Eukaryotes** – cells with a nucleus.

  o Animal cells are eukaryotes.

    ✓ Have a nucleus where genetic material is surrounded by a cell membrane.
    ✓ Cellular and nuclear membrane.
    ✓ Components inside of a cell are called **organelles** and include: mitochondria, Golgi apparatus, ribosomes, endoplasmic reticulum, centrioles, lysosomes, microtubules and vacuoles.

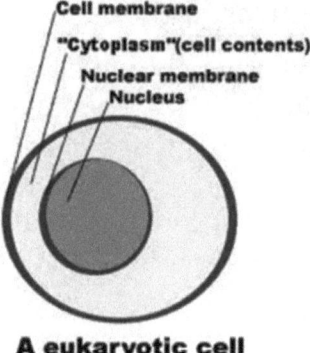

A eukaryotic cell

- Cell or plasma membrane is a double membrane surrounding the cell protoplasm or cytoplasm.

  o Regulates the passage of liquids and gases into and out of the cell.

  o Provides a surface on which chemical reactions take place.

  o Where osmotic activity occurs.

- Protoplasm of living cells is in a colloidal state; made up of medium-sized particles hung in suspension called organelles.

  o Particles are not large enough to settle out but too large to go into solution (i.e., dissolve).

  o Because the particles are smaller, there can be more of them, providing a greater surface area for cellular reactions to take place.

  o They permit reactions to take place rapidly.

  o All cellular metabolic activities take place within the cytoplasm.

  o Cytoplasm is the part between the outer cell membrane and the nuclear membrane.

Chapter Resource

- Nuclear membrane is a double membrane enclosing the nucleus and controlling the movement of materials into and out of the nucleus.

- **Nucleus** is a defined structure within every cell (except red blood cells).
    - Controls activities of the cell.
    - Contains the genetic material responsible for all inherited characteristics.
    - Contains the nucleolus, a smaller body which aids in the synthesis of protein.

- **Endoplasmic reticulum** functions in protein synthesis; two forms: smooth and rough.
    - Provides a transport system between cell parts and a surface on which reactions take place.

- **Ribosomes** are small bodies found on the surface of the endoplasmic reticulum or suspended free of other organelles in the cytoplasm.
    - Site of protein synthesis.

Chapter Resource

- **Mitochondria** are known as the powerhouse of the cell because food is oxidized and energy (ATP) is produced for use in various cellular activities.
    - A fully charged energy unit is called "**ATP**" (adenosine triphosphate).
    - The energy is derived from the sun or from food and can be used when a phosphate group is broken off - making "**ADP**", adenosine diphosphate.
    - Energy-producing reactions in the cell include glycolysis, the citric acid cycle and the electron transport chain.

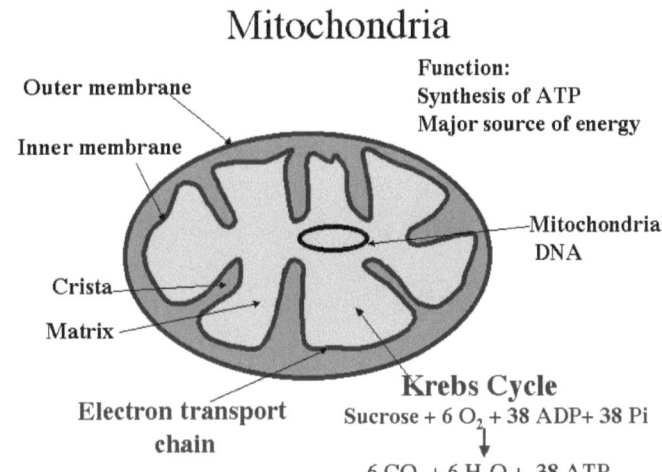

- **Vacuoles** are storage bodies for water, minerals, etc. and can be large in plant cells.

Organelles Unique in Animal Cells

- An **organelle** is a structure or part that is enclosed within its own membrane inside a cell and has a function.

- **Golgi apparatus/bodies** are the site of accumulation for cells that synthesize and secrete lipids and proteins.

- **Lysosomes** are small bodies where large numbers of enzymes are stored.

- **Centrosome** or **centriole** is near the nucleus and functions in cell division.

Organelles Unique in Plant Cells

- Chloroplasts contain chlorophyll and are the site of photosynthesis.

- **Cell wall** (not membrane) is composed of two layers which provide support and protection for the cell.

    - Although somewhat waterproof, the cell wall does not prevent water and substances dissolved in water from passing through.

    - Cell wall is composed of **cellulose**, giving it more rigidity than that of the animal cell membrane.

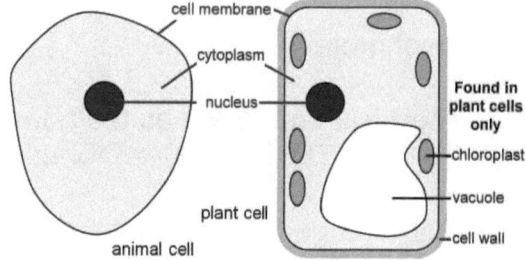

Types of Cells

- Cells **differentiate** (the process by which a less specialized cell becomes a more specialized cell type) into various types depending on their function in the body.

- Types include the following:

    - Absorptive
    - Sensory
    - Nerve
    - Secretory
    - Muscle
    - Reproductive

Chapter Resource

Summary

Components of animal cells are called organelles. The cell or plasma membrane is a double membrane surrounding the cell protoplasm or cytoplasm. The protoplasm of living cells is in a colloidal state; made up of medium-sized particles hung in suspension. The nucleus is a defined structure within every cell. The endoplasmic reticulum is involved in protein synthesis. Ribosomes are small bodies which may be found on the surface of the endoplasmic reticulum (rough) or suspended free of other organelles in the cytoplasm. Mitochondria are known as the powerhouse of the cell. Golgi bodies function in the production of various secretions of the cell. Lysosomes contain digestive enzymes. Cells differentiate into various types depending on function.

Additional Resources

Parker, R. 2013. Equine science. 4th ed. Clifton Park, NY: Delmar Cengage Learning. (Pg. 73-88).

Animal Cells
http://www.biology-pages.info/A/AnimalCells.html

Cells Alive
www.cellsalive.com

Molecular Expressions – Animal Cell Structure
http://micro.magnet.fsu.edu/cells/animalcell.html

Assessment

Take assessment online here: http://tinyurl.com/AnSci-AnimalCells
Download and print the assessment by scanning this QR code or by going to this URL: http://www.tagmydoc.com/Ch09AnSci

10 Cell Functions and Types

Major Concept

Animal cells perform some general functions and some unique functions depending on their type.

Objectives

- Name five types of cells
- List the common functions of cells
- Identify the major components of the cells
- List the functions of the various cell types

Key Terms

- Absorptive cells
- Adipocyte
- Adipose tissue
- Axion
- Calcification
- Cell
- Collagen
- Cytoplasm
- Dentrites
- Diffusion
- Erythrocytes
- Interstitial fluid
- Leukocytes
- Marrow
- Neurons
- Nucleus
- Osmosis
- Osteoblasts
- Protoplasm
- Secretory cells
- Smooth muscle cells
- Striated muscle cells
- Thrombocytes

Chapter Resource

 Complementary *full color* illustrations, photos, charts and graphs are available by scanning this QR code or by following this URL:http://www.tagmydoc.com/AS10 This digital resource will enhance your understanding of the chapter concepts.

What is a Cell?

- A **cell** is a specific, separate mass of living material that is surrounded by a semi-permeable membrane.

    o The basic structural unit of life.

    o All organisms (except viruses) are composed of one or more cells.

    o Three major components of the cell are:

- ✓ **Protoplasm** – the viscid or semi-liquid and jello-like substance which makes up the living cell. It is a catch-all term for all the goods inside the cell membrane.

- ✓ **Cytoplasm** – is the living material inside the cell membrane but outside the nucleus.

- ✓ **Nucleus** – is found in nearly all cells; it contains the hereditary information and is the control center for the cell.

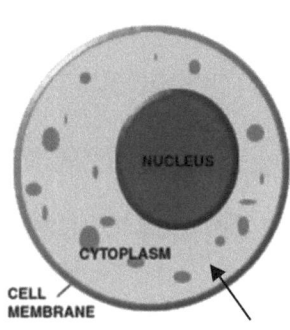

Functions of Cells

- Different cells have different functions in the body.

- What all cells have in common:

Chapter Resource

  o Must take in nutrients to function and must dispose of their waste products. The circulatory system using the blood carries nutrients in and waste out of cells.

  o **Diffusion** – the movement of a substance from a place where it is found in high concentration (relatively large amounts) to a place of low concentration (relatively small amounts). This process continues until the substance is evenly distributed.

  o **Osmosis** – the passage (diffusion) of water across a membrane because of different concentrations on the two sides of the membrane. This happens across a semi-permeable membrane – one which selectively lets some substances across and blocks others.

Types of Cells

- Blood cells – red, white and platelets
- Absorptive
- Secretory
- Nerve
- Sensory
- Reproductive
- Bone
- Fat

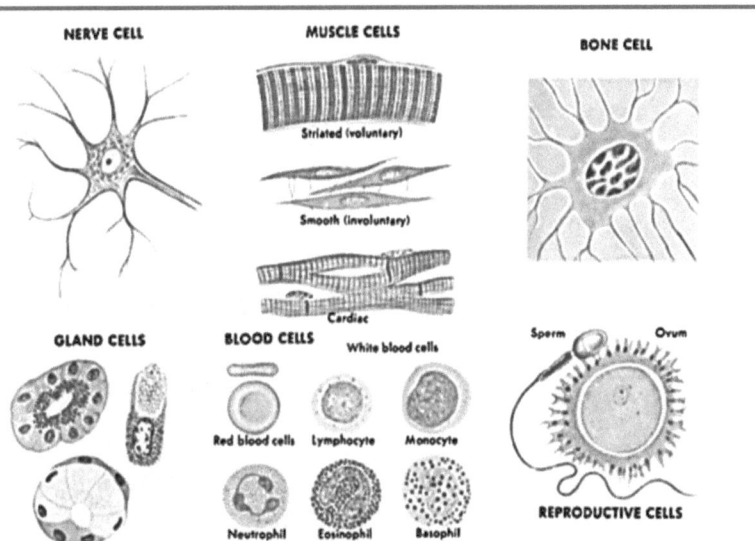

Blood Cells

- Red blood cells or **erythrocytes**

    o An average of 4-5 million red blood cells per milliliter of blood.

    o Formed in bones and contain hemoglobin, whose function is to carry oxygen to cells and carbon dioxide from cells.

    o Soft center of the bone, called the **marrow**, is one place red blood cells are manufactured by the body.

    o Red blood cells are round, with thick edges and a thin center.

    o When they are first formed, they have a nucleus, but it is absorbed before they begin their function in circulation.

    o Red blood cells have a life span of about 120 days in a human body; consequently, the body is continually creating new ones.

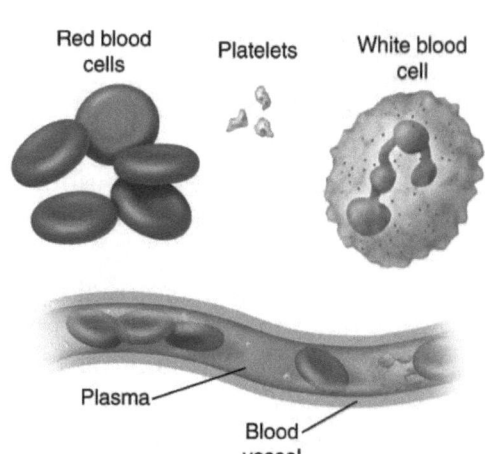

- White blood cells or leukocytes

    o An average of 5-10 thousand per milliliter of blood.

    o Several types, formed in bone marrow and lymph glands. Most are amoeboid, they change their shape and can engulf other cells by flowing around them.

    o White blood cell types include: monocytes, lymphocytes, neutrophils, eosinophils and basophils.

    o Most white blood cells do their work in the **interstitial fluid** (the fluid between the cells).

    o Most spend all their time on body defense:

        ✓ Phagocytic (from the Greek word "phagein" – to eat) **leukocytes** are the amoeboid cells; they flow to the site of infection and engulf the bacteria present.

        ✓ Lymphocytes attack foreign cells directly or secrete an enzyme that immobilizes foreign substances.

        ✓ The way blood cells protect vertebrates against bacteria, foreign substances, etc. is collectively called **immune response**.

- ✓ While defending the body, many white blood cells die. They are the primary component of pus.

- Platelets (**Thrombocytes**)

    - o Third constituent of blood

    - o Do not have a nucleus

    - o A fragment of the cytoplasm

    - o Very important in blood clotting

Chapter Resource

Absorptive Cells

- Occur as continuous sheets on surfaces where material is transported to the cells.

- The single layer of epithelial cells lining the surface of the small intestine selectively absorbs food molecules from the gut into the bloodstream.

- Similar cells are found in the kidney where large surface area is needed for the absorption of protein, water, salts and other materials.

- An absorptive cell needs maximum area for transport, the shape of the cell surface is altered to achieve the optimum transfer of molecules.

Secretory Cells

- Produce products that are subsequently deposited in either the blood stream or a special duct to an organ, where they are used.

- The pancreas and pituitary are glands that have large numbers of secretory cells.

- Proteins and other cell products are synthesized throughout the cytoplasm of these cells and transported to the Golgi apparatus, where they are packaged in membrane-bound vesicles that come to a cell's surface and discharge the secretion outside the cell.

- Secretory cells in the spleen, lymph nodes, and other sites synthesize antibodies for the recognition and destruction of foreign molecules.

Nerve Cells

- A built-in system that works like telephone wires, containing millions of specialized cells called neurons.

    o A **neuron** is a nerve cell which transmits messages from one part of the body to another.

    o A thick telephone cable is made of many single wires bunched together side by side. A single neuron is like one wire in the cable; however, many neurons bunched together like a cable are called nerves.

    o Neurons have the cell parts found in any other cell; however, neurons are very thin and long, do not touch each other and make the sending and receiving of messages (called impulses) faster and more convenient.

    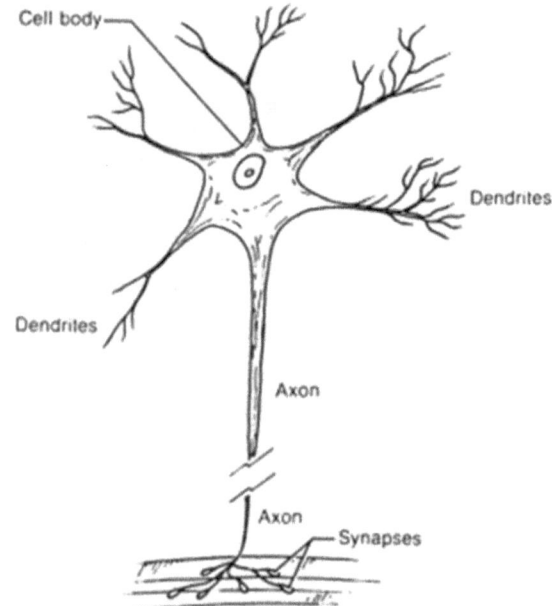

    o A second difference between neurons and other cell types is the branching shape on the ends of the nerve cell. In a typical motor nerve:

    ✓ Branches nearest the cell body (large part with the nucleus) are called **dentrites** and receive the message or stimulus.
    ✓ Branches farthest from the cell body are called terminal buttons or synaptic knobs and send the stimulus on to another nerve or some other type of cell.
    ✓ Size of the nerve cell is also unique. A single motor neuron (with one nucleus) may extend from the spinal cord down the whole length of the leg to the toe.
    ✓ Elongated fiber of the nerve is called the **axion**.

Sensory Cells

- Respond to stimuli by transmitting impulses to the brain.

- Sensory receptors include free nerve endings such as taste buds, hearing receptors and smell receptors.

Muscle Cells

- Three types of muscle cells:

    1. Skeletal
    2. Smooth
    3. Cardiac

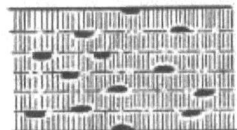  

1. Skeletal   2. Smooth   3. Cardiac

- Skeletal muscle cells

    o Skeletal muscle cells consist of long fibers which show, under the microscope, characteristic cross striations.

    o They have many peripherally located nuclei.

    o They are striated or voluntary muscle cells – ones controlled by conscious choice.

- **Smooth muscle cells**

    o Spindle-shaped cells that are not striated; they contain one centrally located nucleus per cell.

    o Found in the walls of the digestive tract, blood vessels and in the walls of urinary and reproductive organs.

    o Considered involuntary

- Cardiac muscle cells

    o Modified **striated muscle cells** that conduct impulses within the heart much as nerve fibers do in other parts of the body.

Chapter Resource

Reproductive Cells

- Gametes formed after each completion of the process of meiosis, which halves the number of chromosomes in each cell.

- Spermatozoa are motile while ovum are larger and stationary. Fertilization occurs when a sperm is fused with an egg.

Bone Cells

- Bone is a hard connective tissue that makes up most of the skeleton on vertebrates.

  - The outer hard portion is termed **calcification,** making bone unique among tissues.

- Consists of fine-branched cells (called osteoblasts) embedded in a matrix, which the cells secrete.

- Matrix is 30% protein (called **collagen**) and 70% inorganic matter, mainly calcium phosphate.

- **Osteoblasts** are the cells that form layers of bone in the early stages of ossification (bone formation).

- In humans, bone makes up about 7% of the body.

Chapter Resource

Fat Cells

- A single fat cell is referred to as an **adipocyte**; hence, fat is called **adipose tissue.**

- Adipose tissue is deposited around internal organs, between muscle branches and under the skin (and in some individuals between the ears).

- White or yellowish tissue whose primary job is to furnish a supply of reserve energy when food supply is scarce or sporadic.

Summary

Different cells have different functions in the body. Red blood cells are formed in bones and contain hemoglobin, whose function is to carry oxygen to cells and carbon dioxide from cells. White blood cells are formed in bone marrow and lymph glands. Platelets are the third constituent of blood. A neuron is a nerve cell which transmits messages from one part of the body to another. The three types of muscle tissue are: skeletal (striated), smooth and cardiac muscle cells. The different cell types include blood cells – red, white and platelets, absorptive, secretory, nerve, sensory, reproductive, bone and fat cells.

Additional Resources

Parker, R. 2013. Equine science. 4th ed. Clifton Park, NY: Delmar Cengage Learning. (Pg. 75-88).

Cells Alive
www.cellsalive.com

Cell Functions
http://www.biologyjunction.com/cell_functions.htm

Cell Structure and Function
http://biology.unm.edu/ccouncil/Biology_124/Summaries/Cell.html

Function of Animal Cells
http://www.tutorvista.com/biology/function-of-animal-cell

Assessment

Take assessment online here: http://tinyurl.com/AnSci-CellFunctions
Download and print the assessment by scanning this QR code or by going to this URL: http://www.tagmydoc.com/Ch10AnSci

11 Mitosis and Meiosis

Major Concept

Mitosis and meiosis are names for cell division with different purposes.

Objectives

- List, in order, five steps of mitosis
- Identify the differences of mitosis to meiosis

Key Terms

- Anaphase
- Centromere
- Cytoplasm
- Daughter cells
- Furrowing
- Gametes
- Interphase
- Metaphase
- Parent cell
- Prophase
- Somatic
- Telophase
- Tetrad

Chapter Resource

 Complementary *full color* illustrations, photos, charts and graphs are available by scanning this QR code or by following this URL: http://www.tagmydoc.com/AS11 This digital resource will enhance your understanding of the chapter concepts.

Mitosis

- Division of cells in which the genetic material of the cell is duplicated exactly. Cells simply divide and produce daughter cells like themselves. The steps of mitosis include: interphase, prophase, metaphase, anaphase, telophase and return to interphase.

- Step 1: **Interphase**: Resting Stage

    o The period between one division and the next, individual chromosomes are not visible and the nuclear membrane is visible.

    o Although chromosomes cannot be seen, they are present inside the nucleus. Chromosomes are the parts in the nucleus that control inherited traits.

- During the end of this period DNA replicates and each chromosome makes an exact copy of itself. For example, if there are four chromosomes (or two pair) in a resting cell eight are present after copying.

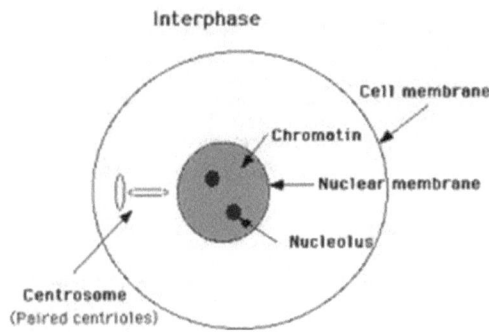

- Step 2: **Prophase**: Preparing to Divide

  - Now two complete sets of chromosomes are joined together in the nucleus.

  - Chromosomes are now coiled tightly and can be seen through a microscope.

  - Nuclear membrane begins to disappear now.

  - For example: given four chromosomes (two pair) a; A; B; b, duplication in prophase creates aa; AA; BB; bb.

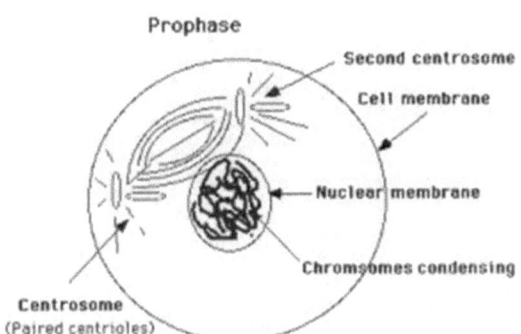

- Step 3: **Metaphase**: Mitosis Starting

  - No longer a division between the nucleus and the cytoplasm.

    ✓ Chromosomes become thicker and shorter and are easier to see through a microscope.

    ✓ Nuclear membrane has faded away.

    ✓ If the cell in Step 1 had four chromosomes four double chromosomes are present. The original and copy appear side by side and are attached to each other by the **centromere**.

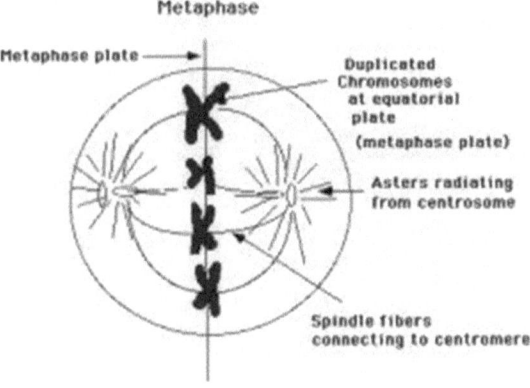

  - Pairs of identical chromosomes line up in the center of the cell. Thin fibers called spindle fibers are attached to the chromosomes.

    ✓ These fibers define the direction in which the chromosomes will later move into the two daughter cells.

- At the end of Step 3, chromosomes are lined up as below:

    ```
    ------a a------
    ------A A------
    ------B B------
    ------b b------
    ```

- Step 4: **Anaphase**: Mitosis Continuing

    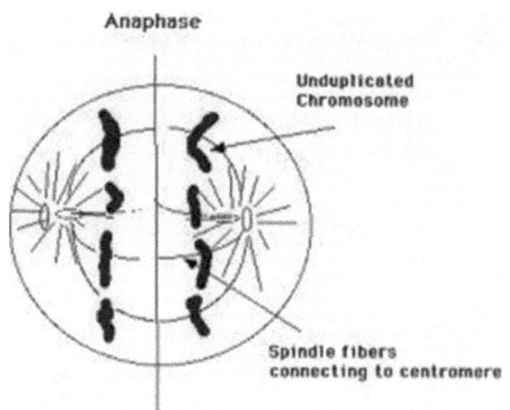

    - Pairs of identical chromosomes separate from one another.

    - Chromosomes move toward the center of each new cell.

        ✓ As one member of each pair moves to one end of the cell, the other member of the pair moves to the other end of the cell creating two groups of identical chromosomes at opposite ends of the cell.

    - Now four, single stranded chromosomes exist at each end of the cell.

    - Diagrammatically the four chromosome pairs will look like this after Step 4:

        ```
        -----a      a------
        -----A      A------   (one set at each end of the cell)
        -----B      B------
        -----b      b------
        ```

    Chapter Resource

- Step 5: **Telophase**: Mitosis Almost Finished

    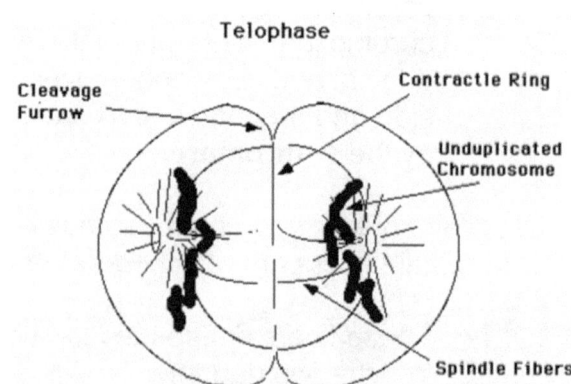

    - A nuclear membrane begins to form around both sets of chromosomes.

    - Chromosomes uncoil and disappear as distinct structures.

    - **Cytoplasm** separates as a new cell membrane (or membrane-like material) forms in the middle of the cell.

    - This is called **furrowing** in animal cells, in which the cell pinches in on all sides until two **daughter cells** are formed.

- o Cell becomes two new daughter cells with identical chromosomes.
    - ✓ New cells are smaller than the original cell from which they came.
    - ✓ Each one will grow and may divide again.
    - ✓ Now two cells and each again have four chromosomes, the same number as the original **parent cell** (a cell that is the source of other cells).
  - o Diagrammatically the two cells after Telophase look like:

    a; A; B; b | a; A; B; b   (two identical cells)

- Step 6: Back to Interphase.

Meiosis

- Cell division process form **gametes** or sex cells.
  - o Involves two successive sequences of cell division and a reduction in the number of chromosomes.
  - o Occurs only in the ovaries and testicles.
  - o Mechanically similar to mitosis.
  - o Described using the same steps as mitosis.

Chapter Resource

- Step 1: Interphase
  - o Similar to mitotic interphase – resting
- Step 2: Prophase
  - o Chromosomes begin to appear as threadlike strands.
    - ✓ Each chromosome has already made a copy of itself while resting and each doubled chromosome has lined up with another chromosome of equal length.
    - ✓ This forms a group of four chromosomes (the likes pair up called a **tetrad**). This forming of groups does not occur in mitosis.
- Step 3: Metaphase
  - o Each group of four chromosomes lines up along the center of the cell.

- o Again, fibers formed in the cell and direct the chromosomes to their correct positions.

- Step 4: Anaphase

    - o Groups of four chromosomes now separate. The original and the copy of each chromosome, however, are still together.

    - o One pair is pulled towards one end of the cell and another pair is pulled to the opposite end of the cell.

- Step 5: Telophase

    - o The cell splits in half, forming two new cells.

    - o At this point, the chromosome number of each new cell is the same as the number in the original cell.

- Step 6: Metaphase

    - o Again, chromosome pairs line up along the center of the cell, but they are lined up in a different direction.

Chapter Resource

- Step 7: Anaphase

    - o Chromosome pairs are now pulled apart. An original moves toward one end of the cell while a copy moves in the opposite direction.

- Step 8: Telophase

    - o Each new cell separates into two; a nuclear membrane begins to form around the chromosome.

    - o Each new cell possesses half the number of chromosomes.

    - o Sex cells (sperm and egg) always have half the number of chromosomes that are present in **somatic** or body cells.

Summary

Mitosis is the division of cells in which the genetic material of the cell is duplicated exactly. Meiosis is cell division that forms sex cells (sperm and egg). In meiosis, two cell divisions occur instead of one. Sex cells always have half the number of chromosomes that are present in somatic or body cells. Meiosis occurs only in the ovaries and testes. The steps for both are interphase, prophase, metaphase, anaphase, telophase and return to interphase.

Additional Resources

Parker, R. 2013. Equine science. 4th ed. Clifton Park, NY: Delmar Cengage Learning.

Meiosis: Where the Sex Starts – Khan Academy
http://tinyurl.com/nx84stg

Mitosis: An Interactive Animation – Cells Alive
www.cellsalive.com/mitosis.htm

Mitosis: Splitting Up Is Complicated – Khan Academy
http://tinyurl.com/mwm2khf

Assessment

Take assessment online here: http://tinyurl.com/AnSci-MitosisandMeiosis
Download and print the assessment by scanning this QR code or by going to this URL: http://www.tagmydoc.com/Ch11AnSci

12 Introduction to Anatomy/Physiology

Major Concept

An understanding of basic animal anatomy and physiology is essential to livestock production.

Objectives

- Identify the four surfaces of the body
- List the nine body systems and their functions

Key Terms

- Abdominal cavity
- Anterior
- Bilateral symmetry
- Caudal
- Comparative anatomy
- Cranial
- Distal
- Dorsal
- Embryology
- Endocrine glands
- Excretion
- Exocrine glands
- Genitalia
- Gross anatomy
- Histology
- Paired structures
- Pelvic cavity
- Pericardium
- Physiology
- Posterior
- Proximal
- Reproductive glands
- Secretion
- Thoracic cavity
- Urogenital system
- Ventral
- Ventral cavity
- Vertebrate

Chapter Resource

 Complementary *full color* illustrations, photos, charts and graphs are available by scanning this QR code or by following this URL: http://www.tagmydoc.com/AS12 This digital resource will enhance your understanding of the chapter concepts.

General Information

- Considerable variation exists among farm animals – cattle, sheep, horses, pigs and poultry.

- To more effectively study varied species with complex body systems, scientists have divided the study of anatomy into several branches which are:

    o **Gross anatomy** is that which can be seen with the naked eye.

- **Histology**, the study of tissue.

  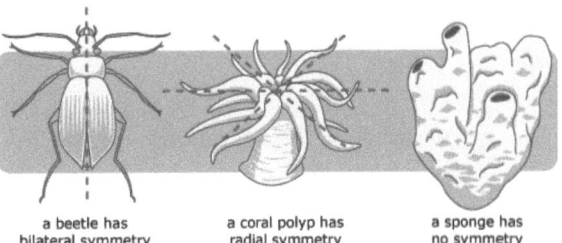
  a beetle has bilateral symmetry   a coral polyp has radial symmetry   a sponge has no symmetry

  ✓ Tissue is an aggregation of cells with similar structure and function; for example, in animal tissues like muscular and nervous tissues; and in plants the meristematic and permanent tissues such as parenchyma and xylem.

- **Comparative anatomy** is the comparison of parts, organs, etc. of different species.

  Chapter Resource

- **Embryology** is the study of body before birth.

  ✓ An important study for animal classification because many of the differences and similarities in various animals are best seen as they are developing before birth.

- **Physiology** is the study of body organ functions, individually or in conjunction with other organs.

• Before studying the internal structure of animals, one should know and be familiar with the external body parts.

Some General Anatomical Terms and Understandings

- **Dorsal** – Upper surface of an animal.

- **Ventral** – Lower or abdominal surface of an animal.

- **Cranial** – Applied to the front or head of an animal; directional terms are anterior and superior.

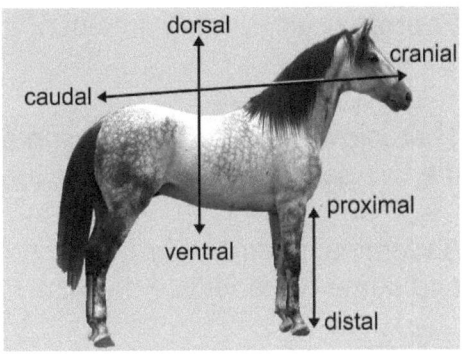

- **Caudal** (posterior) – Pertains to the tail or rear of an animal.

- **Distal** – Refers to the part located the furthest from the center of the body.

- **Proximal** – Situated nearer to the center of the body or point of attachment.

- **Superficial** – Refers to being close to the surface or skin.

- **Bilateral symmetry** – The body is divided into equivalent right and left halves by only one plane.

- **Thoracic cavity** – Chest cavity

- **Pericardium** – Double-walled sac containing the heart and the roots of the great vessels.

- **Abdominal cavity** – Largest body cavity in humans and many animals, and holds the bulk of the viscera.

- **Urogenital** – Of, relating to or denoting both the urinary systems and genital organs.

- An understanding of the general plan of the vertebrate body should include the following:

  o All farm animals are classified as **vertebrates** – having a spinal column.

  o Bodies of vertebrates usually exhibits **bilateral symmetry**:

    Chapter Resource

    ✓ Right and left sides are nearly identical.
    ✓ These similar right and left structures are called **paired structures**, which include ribs, limbs, eyes and most muscles.
    ✓ Exceptions to the rules include the unpaired structures like the tongue, trachea, heart, vertebral column.

- Dorsal part of the vertebrate body contains the brain and the spinal cord.

- **Ventral cavity** contains most of the viscera or guts. (Ventral is the opposite of dorsal.)

- Thoracic (chest area) cavity contains the pericardium which surrounds the heart and the two sacs that surround the lungs.

- Abdominal (stomach area) cavity contains the kidneys, most of the digestive organs, and some reproductive organs.

- **Pelvic** (hips) **cavity** contains the terminal part of the digestive system and all the internal portions of the urogenital system not in the abdominal cavity.

- **Urogenital system** refers to the urinary tract and the accompanying **genitalia** (male and female anatomy).

- **Secretion** is the production of substances useful for the cells in other parts of the body.

- **Excretion** is the expelling of waste products not useful in the animal's body.

- **Endocrine glands** are organs or glands that secrete regulatory substances directly into the circulation and not through a duct.

    o   Important in the control mechanisms metabolism of the body.

    o   Examples are testes, ovaries, adrenals, pituitary and thyroid.

- Reproductive glands include the testes and ovaries.

    o   Produce germ or "sex" cells for reproduction and the hormones testosterone and progesterone.

- **Exocrine glands** have a duct and empty their secretions onto another surface either internally or externally on an epithelial (skin) surface.

    o   Examples are salivary and sweat glands.

- Bodies of all vertebrates are composed of systems (a group of related organs, or other tissues)

Body Systems

- Physiological systems of animals divided into nine systems

    1. Skeletal/Articular
    2. Muscular
    3. Digestive
    4. Urinary
    5. Respiratory
    6. Circulatory
    7. Nervous
    8. Reproductive
    9. Endocrine

    o   Skeletal/Articular is the rigid framework giving the body shape and protecting the internal organs.

      ✓ Consists of bone and cartilage and works in tandem with the muscular system to create movement, includes joints.

    o   Muscular – Provides movement internally and externally; is connected to bones and tendons.

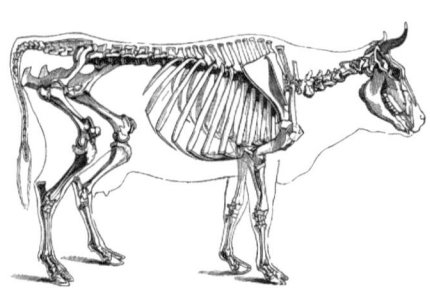

- ✓ Smooth muscle is involuntary; one example is the digestive system.
- ✓ Cardiac muscle is only in the heart and requires no conscious control.
- ✓ Striated or skeletal are voluntary; attaches to parts of bones, tendons and act as sets in either contracting or expanding.

o Digestive system involves all parts of the body in taking food into the body and preparing it for incorporation into the body.

- ✓ Parts include the mouth, pharynx, esophagus, stomach, small intestine and large intestine.

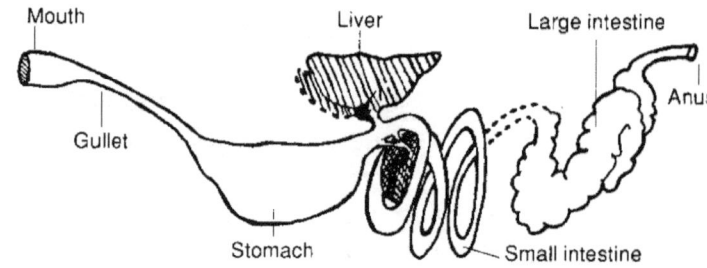

o Urinary system is composed of the kidneys, ureters, bladder and urethra.

- ✓ Processes all the waste products of the life processes.

o Respiratory system takes in oxygen from the environment and delivers it to the tissues and cells of the body; also picks up carbon dioxide from the tissues and cells and delivers it to the environment.

- ✓ Lungs are main component; oxygen is removed by diffusion in the blood.

o Circulatory system distributes blood throughout the body.

- ✓ Consists of heart, veins, arteries and capillaries.
- ✓ Blood carries food substances and oxygen to each cell of the body and takes waste products formed there away from the cells.
- ✓ Blood is the body's regulator and helps to distributes heat, helps regulate the temperature, and neutralize or destroys bacterial and viral invaders.

o Nervous system supplies the body with information about the internal and external environment.

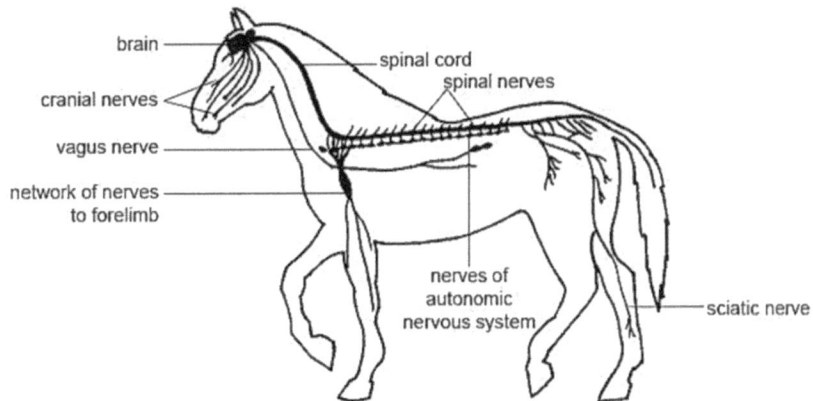

- ✓ System conveys sensation impulses back and forth between the brain or spinal cord and other parts of the body.
- ✓ Main communication system and most important part of the body consisting of the brain, spinal cord, many nerve fibers and sensory receptors.
- ✓ Two main parts – autonomic (automatic) and the central nervous system.
- ✓ Autonomic control the respiratory and digestive systems, eyes, heart and blood vessels, glandular products.
- ✓ Central controls the conscious or voluntary action of the brain stem.

o Reproductive system is the process of creating new organisms of the same species.

- ✓ Produces sperm cells in males and ova or egg cells in females
- ✓ Controlled by hormones.

Chapter Resource

o Provides for the union of male sperm cells and female egg cells, called fertilization, creates a zygote.

o Zygotes develop in the female into a fetus and eventually into a new organism.

o **Endocrine** system consists of ductless glands that produce secretions, hormones that influence the vital functions of the organism from before birth until death.

- ✓ Hormones main purpose is to provide a means of adaptation between the body and its external or internal environment.
- ✓ Many physiological functions are controlled by the endocrine system such as conception, gestation, parturition, digestion, metabolism, growth, puberty and aging.
- ✓ Some of the endocrine glands are pituitary, pineal, thyroid, parathyroid, pancreas, adrenal cortex, adrenal medulla and gonads.

Summary

Some basic terms and knowledge are essential to the study of anatomy and physiology. Descriptive terms useful in the study of anatomy include: cranial, caudal, ventral, distal and proximal. All farm animals are classified as vertebrates. The body of vertebrates usually exhibits bilateral symmetry. Some cells in the glands are specialized for endocrine secretions and others for exocrine secretions. To more effectively study varied species with complex body systems scientists divided the physiological systems of animals into nine systems: skeletal/articular, muscular, digestive, urinary, respiratory, circulatory, nervous, reproductive and endocrine.

Additional Resources

Parker, R. 2013. Equine science. 4th ed. Clifton Park, NY: Delmar Cengage Learning.

Taylor, R.E. and T.G. Field. 2010. Scientific farm animal production: Introduction to animal science. 10th ed. Upper Saddle River, NJ: Prentice Hall.

Assessment

Take assessment online here: http://tinyurl.com/AnSci-IntrotoAnatomy
Download and print the assessment by scanning this QR code or by going to this URL: http://www.tagmydoc.com/Ch12AnSci

13 Comparative Anatomy

Major Concept

Understanding structure, function and adaptation in livestock is essential to management and husbandry.

Objectives

- Define the terms comparative anatomy, homology and analogy
- Identify the role of adaptation to the environment
- Identify the important role mutation can play in the success of a species

Key Terms

- Analogous
- Comparative anatomy
- Homologous

Chapter Resource

Complementary *full color* illustrations, photos, charts and graphs are available by scanning this QR code or by following this URL: http://www.tagmydoc.com/AS13 This digital resource will enhance your understanding of the chapter concepts.

Introduction

- The concept of animal adaptation refers to the genetic and physiological changes that happen in a group of animals because of pressures or changes in the environment.

    o Includes changes in climate (ice age) or feed supply, or pressures from new predators.

- The more an animal adapts or changes to survive in an environment the more successful it will be: faster growth, better reproduction.

Adaptation, Comparative Anatomy, Homology and Analogy

- Study of and comparison of the body parts of different species is called **comparative anatomy**.

    - Comparative anatomy is the study of:
        - ✓ Change, adaptation and mutation.
        - ✓ Invasion of new territory by the species equipped to survive there.
        - ✓ Struggle to be compatible with surroundings that change over time.

- As different species adapt to their changing environments their behavior and their internal and external anatomy changes.

    - Examples:
        - ✓ Forelimb of a human used for manipulating.
        - ✓ Web foot (forelimb) of the seal and whale used for swimming.
        - ✓ Front feet of the mole used for digging.
        - ✓ The wings, posterior (upper or frontal) limb and shoulder spread of the bat.

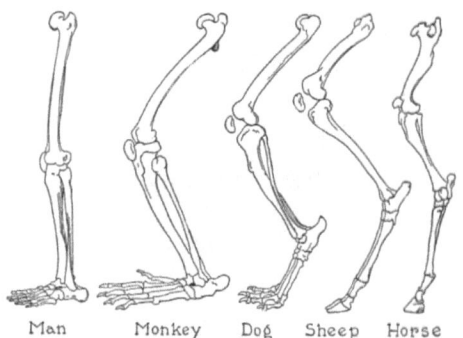

- These body parts and body organs – in different species that have a similar structure, but different uses are called **homologous**.

In Contrast to Homologous is Analogous

- These are animal and plant structures that are similar, but not necessarily homologous. **Analogous** organs are noted for having different structure and origin but are similar in function.

    - A good comparison is that of a butterfly wing to that of a bat wing. They are the same, have the same function, but are different.
        - ✓ The bat wing has an internal skeleton, while the butterfly wing has an external skeleton.

Chapter Resource

- The study of comparative anatomy helps us understand our own history, consider plans for change and improvement of animals of agricultural significance, preserve endangered species and adds to the enlightenment of human kind.

- Darwin believed that homology indicated adaptations that occurred in a common ancestor(s) and that species evolved from those common ancestors as they adapted to their changing surroundings.

- Nobody has seen such gradual changes which occur over millions of years. However, many believe it does suggest the existence of evolution.

Summary

The concept of animal adaptation refers to the genetic and physiological changes that happen in a group of animals because of pressures or changes in the environment. The more an animal adapts or changes to survive in an environment the more successful it will be. The study of and comparison of the body parts of different species is called comparative anatomy. As different species adapt to their changing environments, their behavior and their internal and external anatomy changes. The study of comparative anatomy helps us understand our own history, consider plans for change and improvement of animals of agricultural significance, preserve endangered species and adds to the enlightenment of human kind.

Additional Resources

Comparative Anatomy: What Makes Us Animals?
http://tinyurl.com/k9wnm5j

Inner Body Learning
www.innerbody.com.htm

Assessment

Take assessment online here: http://tinyurl.com/AnSci-ComparativeAnatomy
Download and print the assessment by scanning this QR code or by going to this URL: http://www.tagmydoc.com/Ch13AnSci

14 External Anatomy of Farm Animals

Major Concept

External areas and parts of livestock all have specific names.

Objectives

- Identify the external parts of a dairy cow, horse, pig, goat, sheep, chicken and beef animal.

Key Terms

- Each part or area is a key term

Chapter Resource

 Complementary *full color* illustrations, photos, charts and graphs are available by scanning this QR code or by following this URL: http://www.tagmydoc.com/AS14 This digital resource will enhance your understanding of the chapter concepts.

This area is intentionally left blank.

Parts of a Dairy Cow

- Back
- Barrel
- Bridge of Nose
- Brisket
- Chest
- Chest Floor
- Chine
- Crops
- Dewclaw
- Dewlap
- Ear
- Fore Flank
- Fore Udder
- Fore Udder Attachment
- Forehead
- Heart Girth
- Heel
- Hock
- Hoof
- Hook/Hip
- Horns
- Jaw
- Knee
- Loin
- Mammary Veins
- Milk Wells
- Muzzle
- Neck
- Nostril
- Pastern
- Pinbone
- Point of Elbow
- Point of Shoulder
- Poll
- Rear Flank
- Rear Udder
- Rear Udder Attachment
- Ribs
- Rump
- Shoulder Blade
- Sole
- Stifle
- Switch
- Tail
- Tailhead
- Teats
- Thigh
- Throat
- Thurl
- Withers

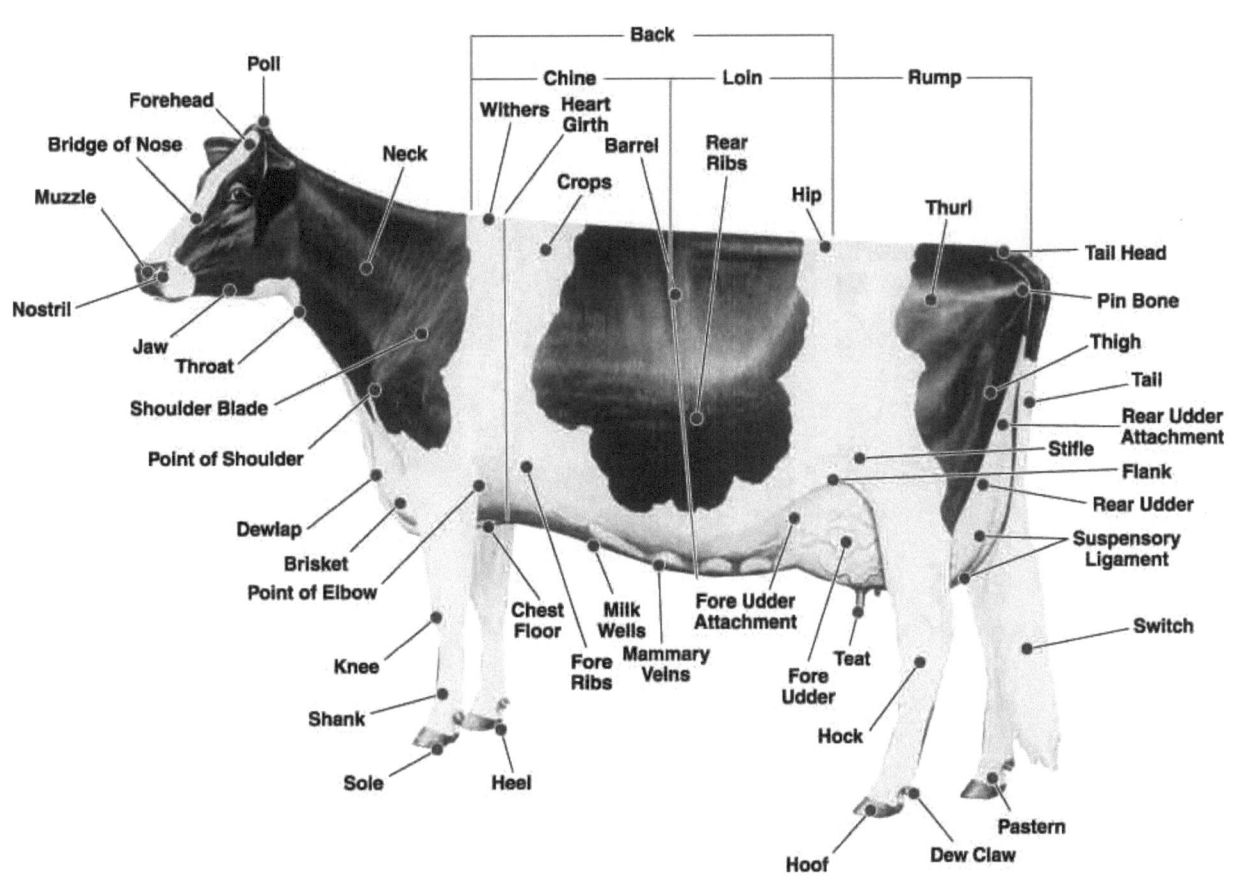

Parts of a Horse

- Back
- Back Tendons
- Belly
- Breast
- Buttock
- Cheek
- Chin
- Chin Groove
- Coronet
- Crest
- Croup
- Dock
- Ear
- Elbow Chestnuts
- Ergots
- Eye
- Fetlock Joint
- Flank
- Flexor Tendons
- Fore Cannon
- Forearm
- Forehead
- Forelock
- Gaskin (Second thigh)
- Girth
- Hallow of Heel
- Hamstring
- Hind Cannon
- Hind Quarters
- Hock
- Hoof
- Jaw
- Jowl
- Jugular Groove
- Knee
- Loins
- Mane
- Muzzle
- Neck
- Nose
- Nostril
- Pastern
- Point of Croup
- Point of Hip
- Point of Hock
- Point of Shoulder
- Poll
- Ribs
- Sheath/Prepuce
- Shoulder
- Stifle Joint
- Tail
- Throat
- Thigh
- Wall of Foot
- Windpipe

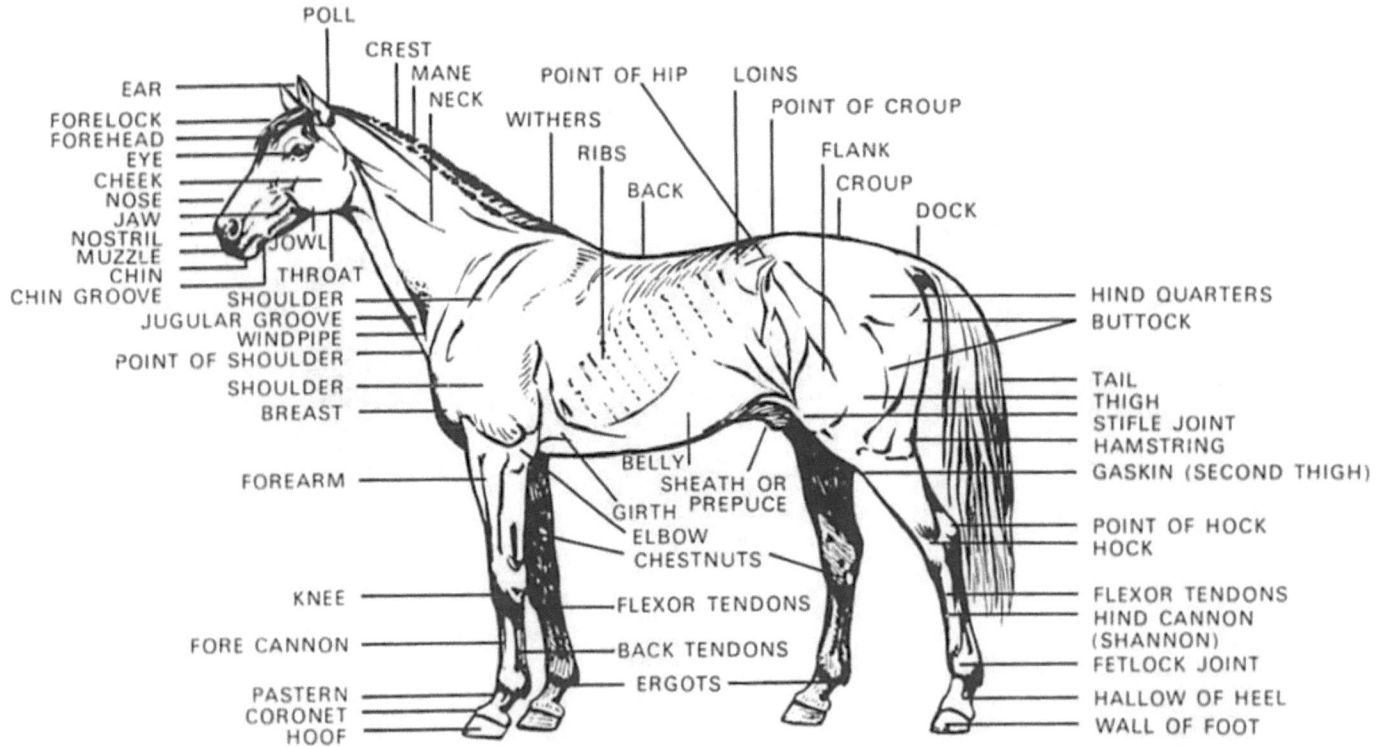

POINTS OF THE HORSE

Parts of a Pig

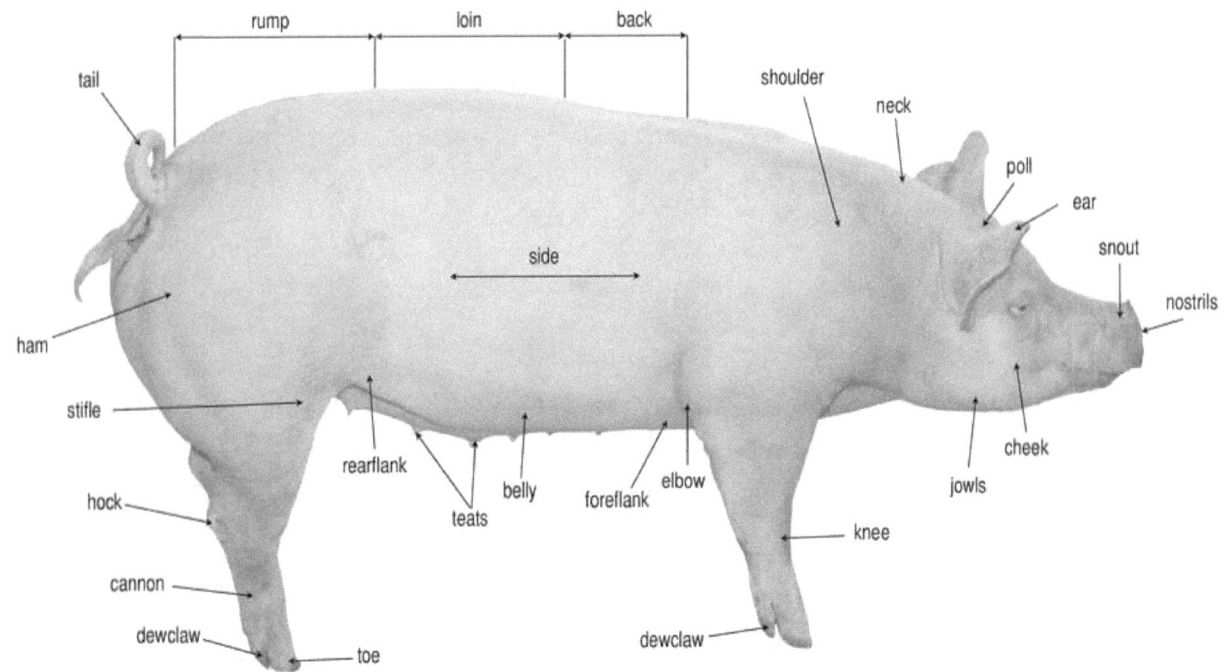

- Back
- Belly
- Cheek
- Dew Claw
- Ear
- Eye
- Face
- Fore Flank
- Foreleg

- Ham
- Hock
- Jowl
- Loin
- Neck
- Pastern
- Poll
- Rear Flank
- Rear Leg

- Rump
- Shoulder
- Side
- Snout
- Tail
- Toes

Parts of a Goat

- Back
- Barrel
- Brisket
- Chest
- Chine
- Crops
- Dewclaws
- Ear
- Elbow
- Eye
- Face
- Fetlocks
- Flank
- Forearm
- Forehead
- Hipbone
- Hock
- Hoofs
- Horn
- Jaw
- Knee
- Loin
- Milk Vein
- Muzzle
- Neck
- Pasterns
- Pin Bone
- Rear Udder Attachment
- Ribs
- Rump
- Shank
- Shoulder
- Tail
- Teats
- Throat
- Thurl
- Udder
- Withers

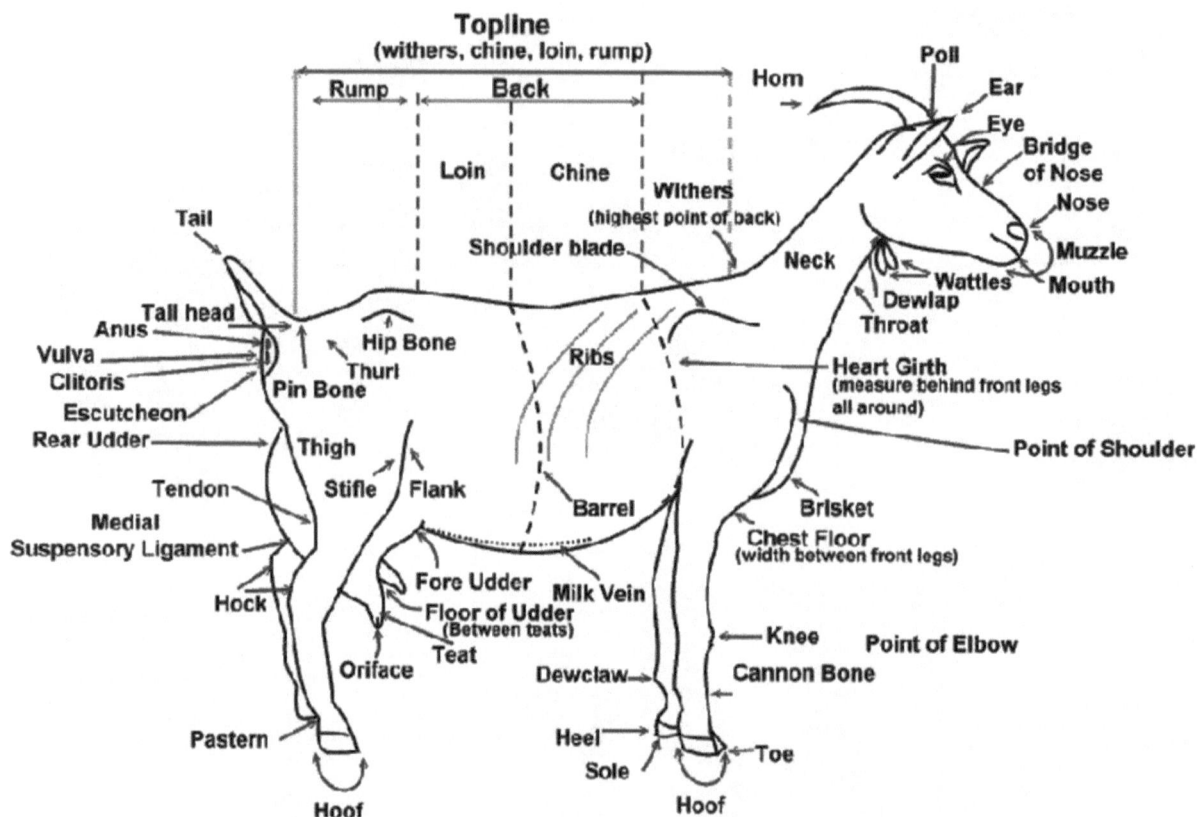

Parts of a Sheep

- Back/Top
- Breast
- Dew Claw
- Dock
- Ear
- Elbow
- Flank
- Fore Arm
- Fore Rib
- Forehead
- Hind Saddle
- Hip
- Hock
- Hoof
- Knee
- Leg
- Loin
- Muzzle
- Neck
- Pastern/Ankle
- Point of Shoulder
- Poll
- Rack
- Ribs/Side
- Rump
- Shoulder
- Throat
- Top of Shoulders
- Twist

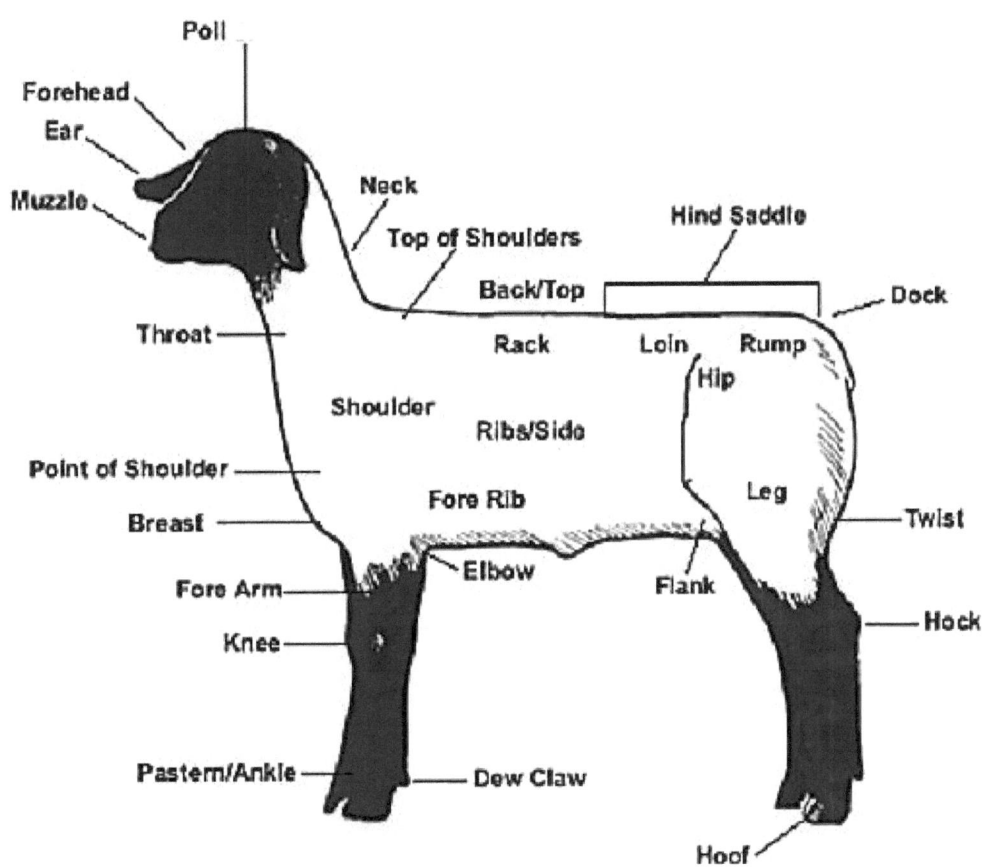

Parts of a Chicken

- Back
- Cape
- Claw
- Comb
- Fluff
- Hackle
- Hock Joint
- Lesser Sickle Feathers
- Main Tail Feathers
- Saddle
- Shank
- Shoulder
- Sickle Feather
- Spur
- Tail Feathers
- Toes
- Wing-Bow
- Wing-Front

Parts of a Chicken

Parts of a Beef Animal

- Back
- Body
- Brisket
- Cannon bone
- Crops
- Dewclaw
- Dewlap
- Elbow
- Face
- Forearm
- Fore flank
- Forehead
- Hind Flank
- Hind Shank
- Hock
- Hoof
- Hooks
- Knee
- Loin
- Muzzle
- Neck
- Pastern
- Poll
- Quarter
- Ribs
- Rump
- Sheath
- Shoulder
- Shoulder vein
- Stifle
- Tailhead
- Underline

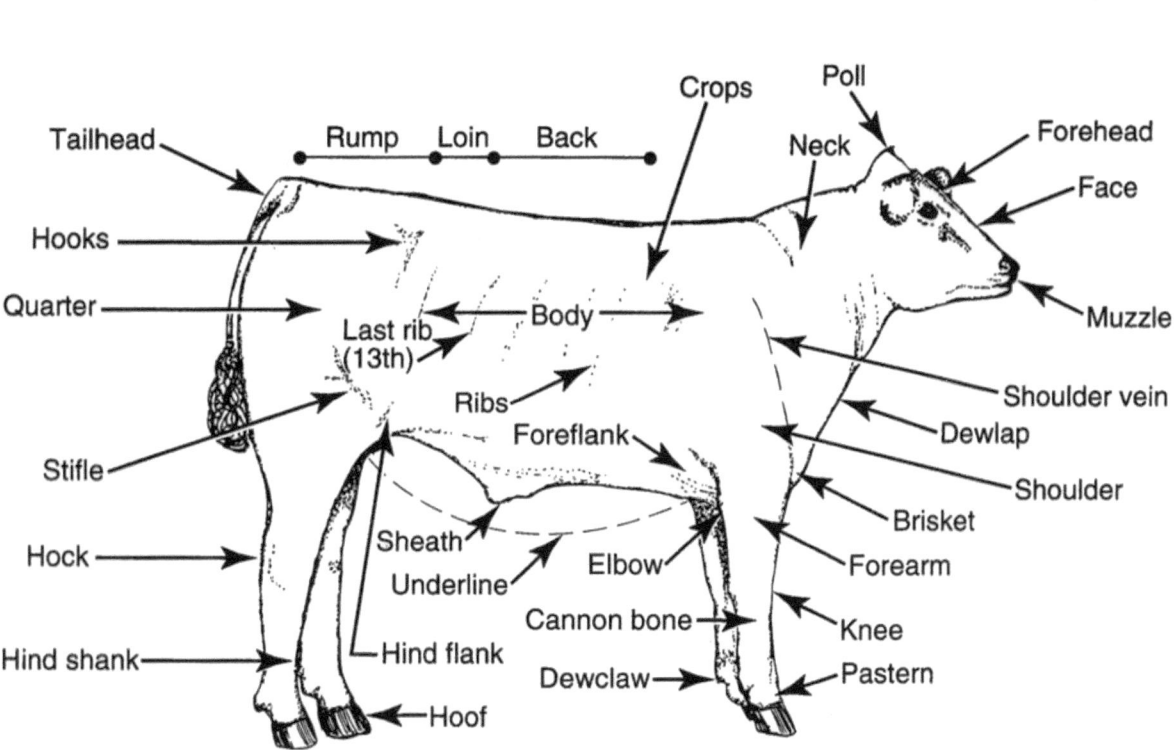

Summary

Much of the external anatomy of livestock is similar for each species but some is unique to a species. To evaluate or judge and to describe disease conditions, livestock producers must be able to locate and name the various parts and areas on the animals.

Additional Resources

Taylor, R.E. and T.G. Field. 2011. Scientific farm animal production. 10th ed. Upper Saddle River, NJ: Prentice Hall.

External Anatomy of Animals
http://tinyurl.com/nxsklrb

Assessment

Take the assessment online here: http://tinyurl.com/AnSci-ExternalAnatomy
Download and print the assessment by scanning this QR code or by going to this URL: http://www.tagmydoc.com/Ch14AnSci

15 Skeletal System

Major Concept

Skeletal system provides a framework of hard tissue that support and act as levers for movement of the body and to protect soft tissues.

Objectives

- Identify the bones in the axial and appendicular skeleton
- List the functions of bones in the axial and appendicular skeleton
- Name the divisions of the spinal/vertebral column
- List the components of movable joints

Key Terms

- Appendicular skeleton
- Axial skeleton
- Coccygeal
- Ligaments
- Lumbar group
- Mandible
- Metacarpal bones
- Nasal cavity
- Phalanges
- Sinuses
- Synovial fluid
- Temporal bone
- Tendons
- Thoracic limbs
- Tibia
- Turbinates

Chapter Resource

Complementary *full color* illustrations, photos, charts and graphs are available by scanning this QR code or by following this URL: http://www.tagmydoc.com/AS15 This digital resource will enhance your understanding of the chapter concepts.

Bone Structure

- Compact and contains blood vessels, nerves and marrow.

- Can repair itself and can bend.

- Organic material gives it elasticity and inorganic material gives it rigidity.

- Bone is composed of 1/3 organic and 2/3 inorganic material.

Skeletal System

- Skeleton is the framework of hard tissue which gives the body support and protects soft tissues.

- Serves as levers in movement.

- Number of bones in the skeleton varies with the age and species of the animal.

- Bones may be described as long (femur), flat (skull), short (vertebrae) and irregular (bones of the pelvis).

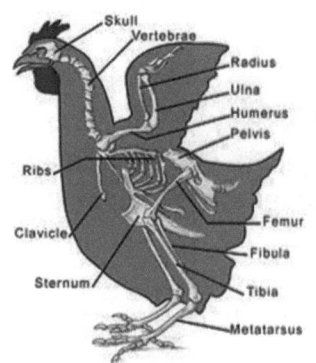

Axial Skeleton

- Axial skeleton refers to the bones of the skull, vertebral column, ribs and sternum.

- Skull includes the following:

    o **Nasal cavity** – The cavity in which the olfactory organs of vertebrate animals are located.

    o **Mandible** – Lower jaw or jawbone.

    o **Temporal bone** – Either of a pair of compound bones forming the sides and base of the skull.

    o **Sinuses** – Hollow walled spaces.

Equine

    o **Turbinates** – Cartilaginous bone (not hard) covered by highly vascular (many blood vessels) mucosa which serves to clean and warm the air as the animal breathes in.

- Vertebral/spinal column includes the cervical, thoracic, lumbar, sacral and **coccygeal** (referring to the coccyx, the small tail-like bone at the bottom of the spine) sections.

    o Protects the spinal cord which runs up the hollow middle of the vertebra.

    o Some vertebra is moveable, others are not.

    o Serves as an attachment for the appendicular skeleton – examples include:

        ✓ Ribs which attach to the spinal column

- ✓ **Thoracic limbs** – arms or front legs (including the scapula, arm, radius, ulna, manus, carpus and digits)
- ✓ Pelvic limbs (legs or hind legs)

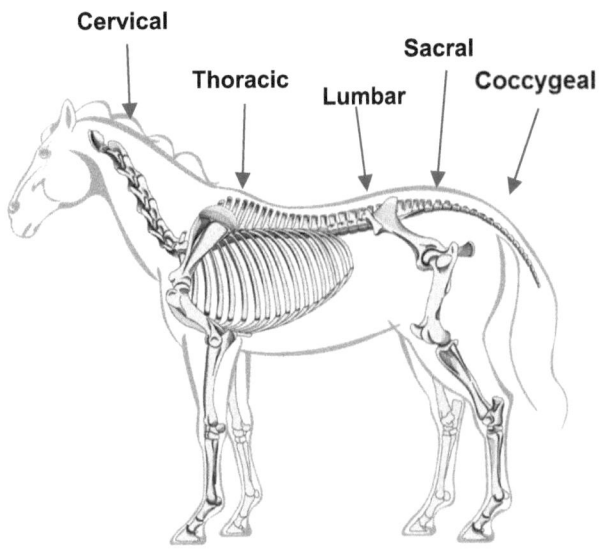

o Cervical or neck area
  - ✓ Involved with head and neck movement.
  - ✓ Most flexible part of the axial skeleton.

o Thoracic Area (shoulders)
  - ✓ Forms the upper wall of the chest cavity with the 18 ribs attached.
  - ✓ Peaks at the summit of the withers.
  - ✓ Has limited movement and flexibility.

o **Lumbar group** (lower back)
  - ✓ Usually includes six vertebrae (Arabians may have 5).
  - ✓ Provides the framework for the loin area.
  - ✓ Greater flexibility than in the thoracic group but less than the cervical group.

Chapter Resource

o Sacral group (rump)
  - ✓ Includes five vertebrae fused into one bone (sacrum).
  - ✓ Highest point of the croup.
  - ✓ Connected by a firm joint with the hip (pelvis) bones on each side of the sacrum.

o **Coccygeal** (tail)
  - ✓ 15-21 coccygeal vertebrae.
  - ✓ Essentially no spinal canal through the middle of the vertebrae.

Appendicular Skeleton

- Appendicular skeleton refers to the limbs.

    o Used for locomotion and can also be used for eating and defense.

    o Connects with the axial skeleton by muscles and in the case of the hind leg, by bony joints and muscles.

- Foreleg includes:

    o Scapula (shoulder blade) which is flat and triangular in shape.

    o Humerus that extends down and back from the scapula.

    o Radius and ulna which are fused together and connect with the humerus and form the elbow joint.

    o Carpal bones

        ✓ Knee bones

        ✓ Serve to absorb and dissipate concussive action (the pounding of the hooves as the animal runs).

    o **Metacarpal bones** (three) – which extend from the knee to fetlock.

    o **Phalanges** – which corresponds to the hand of man.

    o Sesamoids – forms part of the fetlock joint.

    o Distal or navicular bone – located at the rear of the coffin bone.

- Hindleg includes:

    o The femur, largest bone of the upper leg; corresponds to thigh bone of man.

    o Patella

        ✓ Small bone in front of the stifle joint.

        ✓ Corresponds with the kneecap of man.

    o **Tibia** and fibula

        ✓ Corresponds with the shin bone of man.

Chapter Resource

> ✓ Starts at the stifle joint, extends downward and backward to the hock joint.

- o Tarsus (hock)

- o Metatarsus or cannon bone of the hindleg.

- o Phalanges and sesamoids of the hindlegs that correspond to those of the foreleg.

Articulation of the Joints

- Immovable joints are those with no intervening tissue between bones (skull).

- Freely moveable joints are those that can move in several directions like those of the shoulders and hips.

- Cavity of the joint (hollow pocket) may contain **synovial fluid**, which is secreted by the synovial membrane. The fluid helps lubricate the joint.

- **Ligaments** are strong white fibrous tissues that connect bone to bone.

- **Tendons** are dense cord-like tissues that connect muscles to bones.

Summary

The skeleton is the framework of hard tissue which gives the body support and protects soft tissues. Bone is compact and contains vessels, nerves and marrow. It can repair itself and can bend. The skull includes the nasal cavity, mandible, temporal bone, sinuses - hollow walled spaces and turbinates. The vertebral column includes the cervical, thoracic, lumbar, sacral, and coccygeal sections. Immovable joints are those with no intervening tissue between the bones. The skeletal system is the framework of hard structures which supports and protects the soft tissue. Bone is composed of 1/3 organic and 2/3 inorganic material. The axial skeleton includes the skull, spinal column, cervical area, thoracic area, lumbar group, sacral group and the coccygeal. The appendicular skeleton contains the scapula, humerus, radius and ulna, carpal bones, metacarpal bones, phalanges, sesamoids, distal or navicular bones, femur, patella, tibia and fibula, tarsus, metatarsus and phalanges and sesamoids.

Additional Resources

Taylor, R.E. and T.G. Field. 2011. Scientific farm animal production. 10th ed. Upper Saddle River, NJ: Prentice Hall.

The Skeletal System: It's Alive!
https://www.youtube.com/watch?v=RW46rQKWa-g

WikiBooks: Anatomy and Physiology of Animals – The Skeleton
http://en.wikibooks.org/wiki/Anatomy_and_Physiology_of_Animals/The_Skeleton

Assessment

 Take assessment online here: http://tinyurl.com/AnSci-SkeletalSystem
Download and print the assessment by scanning this QR code or by going to this URL: https://www.tagmydoc.com/Ch15AnSci

16 Digestive System

Major Concept

A basic understanding of the digestive system aids the producer in nutrition and management.

Objectives

- List the functions of the digestive organs in various livestock
- Name the structures/organs in the digestive systems of the horse, pig, cow and chicken
- Identify the differences between the digestive systems of the horse, pig, chicken and cow

Key Terms

- Abomasum
- Alimentary canal
- Cardiac
- Catalysts
- Cecal fermenters
- Cecum
- Complex stomach
- Crop
- Digesta
- Esophagus
- Fundus
- Gastric
- Gizzard
- Hydrolyze
- Ingesta
- Mastication
- Mouth
- Omasum
- Papillae
- Peristaltic movement
- Pharynx
- Proventriculus
- Pylorus
- Reticulum
- Ruminants
- Rumination
- Salivary glands
- Tongue
- Villi

Chapter Resource

Complementary *full color* illustrations, photos, charts and graphs are available by scanning this QR code or by following this URL: http://www.tagmydoc.com/AS16 This digital resource will enhance your understanding of the chapter concepts.

System Components

- General Description: The digestive system is composed of organs that are involved in the process of digestion of food.

- Digestion – Process in alimentary canal by which feed is broken down mechanically and enzymatically and converted into simpler chemical compounds suitable for absorption and assimilation into the body.

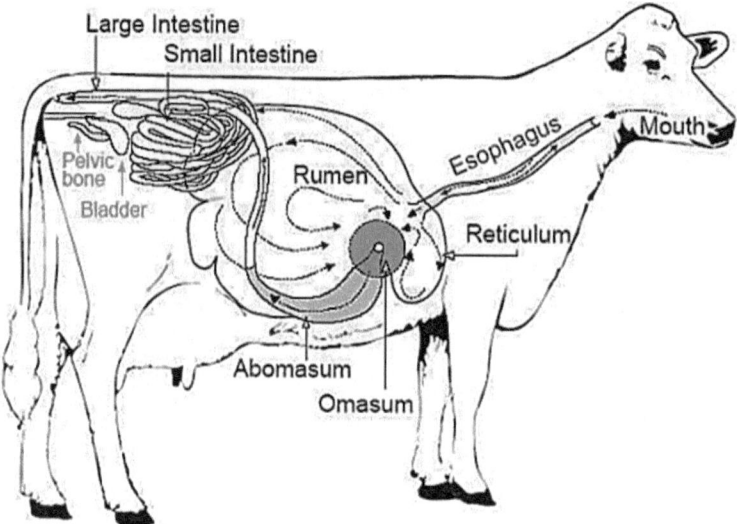

- **Alimentary canal** is a long winding tube with various enlarged sacs, beginning with the mouth and ending at the anus.

  - Food undergoes complex mechanical and chemical changes as it moves through the alimentary canal or tube.

  - Humans and pigs have a simple stomach with an extensive intestinal system.

  - **Ruminants** such as cattle, sheep and goats have a large compartmented and very **complex stomach** with a less simple intestinal system.

  - Horses and rabbits have a simple stomach with an extensive intestinal system with an enlarged cecum.

- **Cecum** acts like a rumen and is involved with microbial digestion (fermentation).

- **Mouth** is the initial opening of the alimentary canal.

  - **Tongue** is a tool of prehension that is used to grasp the food or to guide it in the mouth and on to the throat.

  - **Salivary glands** (exocrine glands) secrete juices in the mouth that are mixed with the food.

    ✓ These juices contain enzymes that start carbohydrate digestion.

    ✓ Moistening also makes the food easier to swallow.

- o **Mastication** is the term used to describe the chewing and crushing of food in preparation for swallowing.
    - ✓ More complex in herbivores than in carnivores (animals that feed on flesh).
- o **Pharynx** is a short, funnel shaped muscular sac between the mouth and esophagus.
    - ✓ Part of the digestive and respiratory tracts.
    - ✓ Food passes through the pharynx into the esophagus by muscle action.
- **Esophagus** is a muscular tube that connects the pharynx to the stomach.
    - o Muscular contractions move the food down the esophagus to the stomach.
- Stomach is in the abdominal cavity between the esophagus and small intestine.
    - o Simple stomach
        - ✓ Humans, swine, rabbits and horses have a simple stomach.
        - ✓ Divided into three regions; **cardiac**, **fundus** and **pylorus**.

        Chapter Resource

    - o Digestion in the simple stomach is:
        - ✓ Mechanical (physical), through chewing and muscular contractions.
        - ✓ Chemical, through gastric secretions of enzymes which soften food and begin the breakdown of the nutrient molecules. Enzymes are **catalysts** – they start chemical reactions.
        - ✓ Protein is the primary food component broken down in the stomach.
    - o Three primary sources of enzymes which break down fats, proteins and carbohydrates include:
        1. **Gastric enzymes** – Enzymes that are secreted in the stomach.
        2. Liver and pancreatic – Mostly fats, in the small intestine (duodenum - first part).
        3. Intestinal – Mostly carbohydrates and some proteins, in the small intestine (where digestion is completed).
    - o Ruminant stomach
        - ✓ Sheep, cows and goats are all examples of ruminants.

- ✓ Ruminant stomach occupies 3/4 of the abdominal cavity, mainly on the left side.

  o Composed of four compartments:

    1. Rumen
    2. Reticulum
    3. Omasum
    4. Abomasum

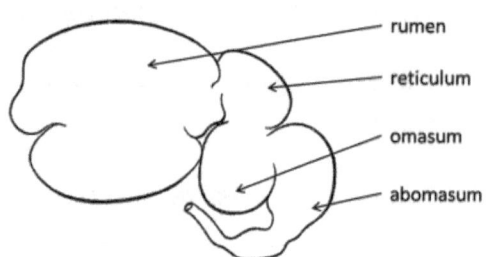

  - ✓ Rumen or paunch is a mucous membrane studded with **papillae** (any small, nipplelike process or projection) to increase surface area.
  - ✓ In mature animals, it is 80% of bovine ruminant stomach system, 30% in young.
  - ✓ Reticulum (honeycomb) composes about 5% of the bovine stomach.
  - ✓ It has a raised mucosa (1/2" high) for increased absorptive area; is located near the heart; and prevents large indigestible objects from entering the stomach (nicknamed the hardware stomach).
  - ✓ **Omasum** (many-plies) is 7-8% of the bovine ruminant stomach and absorbs mostly water.
  - ✓ **Abomasum** is the true stomach.
  - ✓ In mature animals, it composes 7-8% of the area and in young animals about 70%. (The difference is that young animals have not yet developed the rumen and are still living on milk which would spoil in a fully developed rumen; however, it is perfect for a simple stomach or its substitute in ruminants, the abomasum).

Chapter Resource

- Ruminant digestion

  o Heavy items settle in the **reticulum** which is open to the rumen.

  o **Digesta** (partially digested food) can pass back and forth from the reticulum to the rumen.

    - ✓ Lighter food collects in the rumen.

  o Gas from microbial activities collects above the rumen content and is absorbed. To function effectively the rumen requires a lot of water (provided by saliva).

    - ✓ Omasum absorbs most of the excess water from the digesta as it passes from the rumen.

- ✓ Abomasum is the true stomach with typical enzyme action. The food remains here for a relatively short time compared with the time in the rumen.
  - o **Rumination** is the regurgitation, remastication, resalivation and reswallowing of food.
    - ✓ Purpose is to further smash and break up the less digestible parts of food.
    - ✓ This breakdown gives more surface area for the bacteria to work on the cellulose (less digestible plant part) before it goes down the line.
    - ✓ Resalivation helps provide the water needed for rumen (and bacterial) action.

- Small intestine is a long coiled tube connecting the stomach with the large intestine.
  - o Food moves through the intestines by muscular contractions called **peristaltic movement**.
  - o **Ingesta** (substances taken into the body as nourishment) are further mixed with digestive enzymes and the rest of the breakdown and absorption of nutrients occurs here.
  - o To improve absorption, the surface area is increased with **villi**.

- Large intestine is considered to include the cecum, colon and rectum.
  - o It absorbs the water and the residue (feces) becomes more solid.
  - o Other than absorption of some vitamins and minerals, very little nutrient absorption takes place here.
  - o Exceptions of course are the **cecal fermenters** meaning that they digest nutrients by means of the cecum rather than by a multi-chambered stomach such as the rabbit and horse.

- Poultry digestion

  Chapter Resource

  - o **Crop** – an expanded, muscular pouch near the gullet or throat.
  - o **Proventriculus** – acts as the true stomach of a bird.
  - o **Gizzard** – an organ found in the digestive tract of a chicken.

- Accessory organs:
  - o Pancreas secretes enzymes which **hydrolyze** (decompose by reacting with water) starch and breaks down proteins.

- o Liver is the largest internal organ and secretes bile, which helps digest fats.
    - ✓ Bile is stored in the gall bladder.
    - ✓ Liver is a very complex organ with many other functions in the body including storing iron, handling fats and carbohydrates in the blood.

Summary

The digestive system is composed of the organs involved in the process of digestion of food. The mouth is the initial opening of the alimentary canal. The pharynx is a short, funnel shaped muscular sac between the mouth and esophagus. The esophagus is muscular tube that connects the pharynx to the stomach. The stomach is in the abdominal cavity between the esophagus and small intestine. The small intestine is a long coiled tube connecting the stomach with the large intestine. The large intestine is considered to include the cecum, colon and rectum.

Additional Resources

Taylor, R.E. and T.G. Field. 2011. Scientific farm animal production 10th ed. Upper Saddle River, NJ: Prentice Hall.

Animal Sciences - Purdue
https://ag.purdue.edu/ansc/Pages/default.aspx

Avian Digestive System
http://articles.extension.org/pages/65376/avian-digestive-system

The Digestive Tract of the Pig
http://edis.ifas.ufl.edu/AN012

Assessment

Take assessment online here: http://tinyurl.com/AnSci-DigestiveSystem
Download and print the assessment by scanning this QR code or by going to this URL: https://www.tagmydoc.com/Ch16AnSci

17 Hormone Function

Major Concept

Hormones from the endocrine glands control many functions in the animal body.

Objectives

- Define the term endocrinology
- List examples of the kinds of effects that hormones have on the body
- Name five hormones, their action and the gland that secretes them
- Identify the role of the pituitary and gonads on governing sexual development in animals

Key Terms

- Endocrinology
- Follicles
- Gonadotropins
- Homeostatic mechanism
- Secondary sex characteristics
- Testosterone

Chapter Resources

Complementary *full color* illustrations, photos, charts and graphs are available by scanning this QR code or by following this URL: http://www.tagmydoc.com/AS17 This digital resource will enhance your understanding of the chapter concepts.

Endocrinology

- Science that deals with the study of the endocrine glands and their secretions, the hormones.

  o Endocrine glands are small and are located throughout the body. They contain cells that secrete the hormones. They are:

   ✓ Ductless
   ✓ Release their secretions directly into the bloodstream, rather than through a tube to another organ.

  o Once the hormones are in the bloodstream they are carried to the various regions of the body where they cause certain organs to perform specific functions. Hormones have an important effect upon body function.

- - Examples include:
    - ✓ Growth and fattening
    - ✓ Reproduction
    - ✓ Lactation
    - ✓ Egg laying

  Chapter Resource

  - Except for small amounts that may be held in the endocrine organs themselves, hormones are not stored in the body.
  - To correct a hormone deficiency, it is necessary to give repeated small doses rather than one large dose as is used for treating some other animal ailments.

- Factors related to the functions of hormones include:
  - Only small amounts are required for proper function.
  - Affect growth, body shape, and the way the body uses its food and help the body to adjust to changes in the environment.
  - An overdose of hormones can cause more problems than solutions, so treatment must be careful and measured.
  - Individuals may react differently to the same hormone.
  - A given hormone may act differently on the tissues of different species.
  - Hormones appear to regulate biochemical reactions but do not initiate them.

- For good health and development, the glands must work together as a unit.
  - If a gland should become either too active or not active enough, illness results.
    - ✓ Examples include – too active: hyperthyroidism, giantism.
    - ✓ Not active enough: dwarfism and diabetes.
  - Because the activity of the endocrine glands affects the complete animal system, all hormonal activities are carefully regulated by the body's **homeostatic mechanism** (physiologic control system).

Location of Endocrine Glands

- Location similar in other livestock

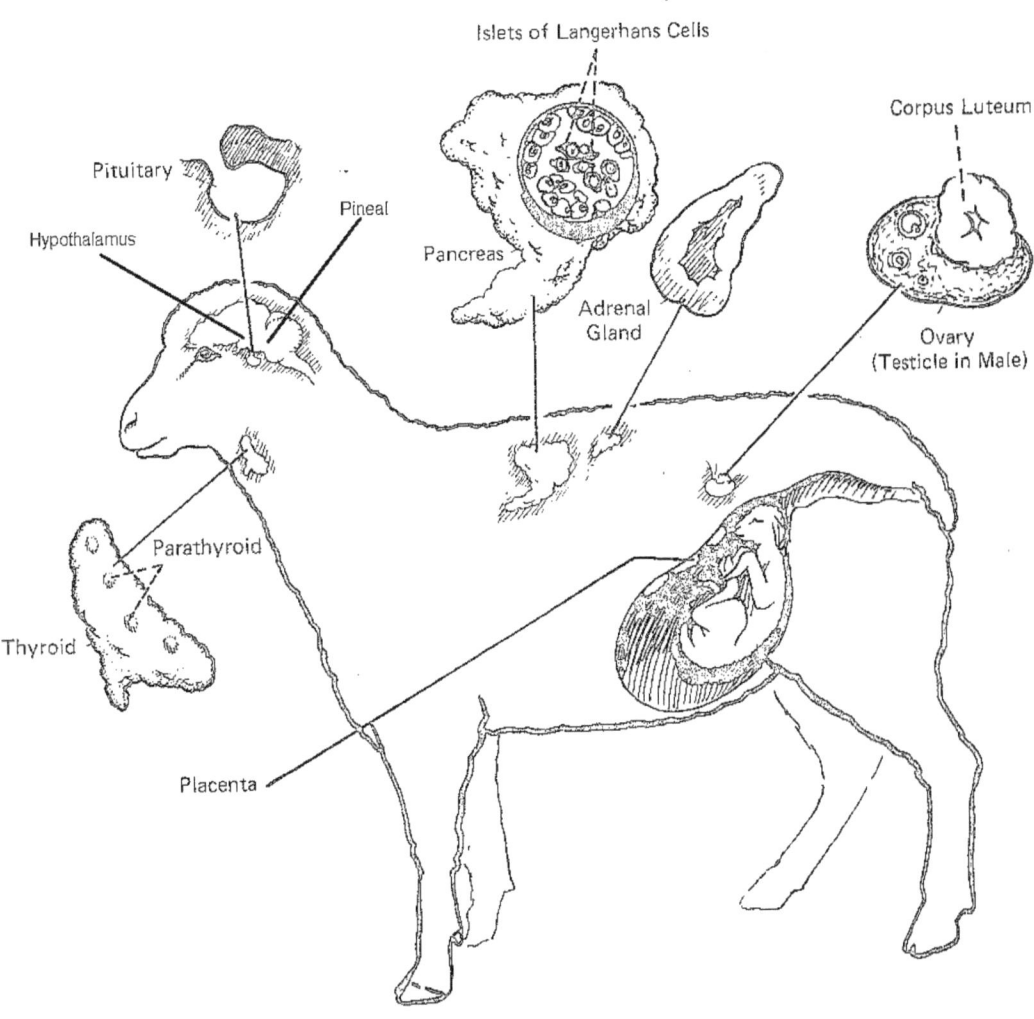

Endocrine Glands, Hormones and Functions

Gland	Hormone	Function
Hypothalamus	Releasing hormones	Controls the pituitary gland
Posterior pituitary	Oxytocin	Stimulates uterine contractions and milk letdown
Posterior pituitary	Vasopressin or ADH	Increases water absorption in kidney
Anterior pituitary	Growth hormone (STH)	Promotes growth of most tissues
Anterior pituitary	Prolactin (LTH)	Promotes lactation; stimulates corpus luteum
Anterior pituitary	Adrenocorticotropic hormone (ACTH)	Stimulates adrenal cortex
Anterior pituitary	Thyroid-stimulating hormone (TSH)	Stimulates thyroid gland
Anterior pituitary	Follicle-stimulating hormone (FSH)	Stimulates follicle growth on the ovaries and sperm production in the male

Gland	Hormone	Function
	Luteinizing hormone (LH)/ Interstitial cell-stimulating hormone (ICSH)	LH stimulates ovulation, corpus luteum function, secretion of progesterone, and secretion of estrogen in the female; ICSH facilitates production of testosterone in the male
Pineal	Melatonin	Aids in adaptation to light-dark cycles
Thyroid	Thyroxine	Controls metabolism and affects growth, reproduction, and nutrient assimilation
	Thyrocalcitonin	Decreases blood serum levels of calcium
Parathyroid	Parathormone	Regulates metabolism of calcium and phosphorus
Pancreas	Insulin and glucagon	Regulate glucose metabolism
Adrenal cortex	Glucocorticoids	Stimulate conversion of protein to carbohydrates for energy; decrease inflammation and immune response
	Androgens	Regulate masculine secondary sexual characteristics
	Mineralocorticoids	Regulate sodium and potassium metabolism
Adrenal medulla	Epinephrine and norepinephrine	Prepare animal for emergencies; mobilize energy
Testes	Testosterone	Develops and maintains accessory sex glands; stimulates secondary sexual characteristics, regulates sexual behavior and sperm production
Ovary	Estrogen	Promotes female sexual behavior; stimulates secondary sexual characteristics, growth of reproductive tract, mammary growth, and feedback control
	Progesterone	Prepares uterus, maintains pregnancy and prepares mammary glands for lactation, and provides feedback control
	Relaxin	Facilitates dilation of birth canal
Gastrointestinal tract	Secretin, enterokinin, cholecystokinin, enterogastrone	Control secretions and motility of digestive tract

Study of Hormone Function in Males

- Reproduction in males is controlled by at least three hormones, two from the pituitary gland and one from the testes.

    o Pituitary effects and/or controls the activities of many other glands through its own secretions.

    o Testes are the glands in the male that are specifically related to maleness.

       ✓ Traits such as aggressiveness, size, etc. generally grouped under the category of **secondary sex characteristics** or those that begin to show with the onset of puberty.

- Two hormones from the pituitary are called **gonadotropins** because they stimulate the gonads or sex glands – in this case the testes.

- o Released in response to hormones from the hypothalamus.

- o First, follicle stimulating hormone (FSH). In the female, it stimulates the production of **follicles**; in the male, it stimulates the seminiferous tubules to produce sperm.

- o Second, luteinizing hormone (LH) which stimulates the interstitial cells of the testes (cells of Leydig) to produce the male hormone testosterone.

- **Testosterone** is the hormone produced in the male gonads or testes.

  - o Causes the growth, development and secretory activity of the male reproductive organs and the accessory sex glands.

  - o Needed for the survival of spermatozoa.

  - o Governs the development of secondary sex characteristics such as the crest, male voice, sex drive, etc.

Chapter Resource

Summary

Endocrinology is the science that deals with the study of the endocrine glands and their secretions, the hormones. Endocrine glands are small and are located throughout the body. Once the hormones are in the bloodstream they are carried to the various regions of the body where they cause certain organs to perform specific functions. Hormones have an important effect upon many body functions.

Additional Resources

Taylor, R.E. and T.G. Field. 2011. Scientific farm animal production. 10th ed. Upper Saddle River, NJ: Prentice Hall.

WikiBooks: Anatomy and Physiology of Animals - The Endocrine System
http://en.wikibooks.org/wiki/Anatomy_and_Physiology_of_Animals/Endocrine_System

WikiVet: Anatomy and Physiology - Endocrine System
http://en.wikivet.net/Category:Endocrine_System_-_Anatomy_%26_Physiology

Assessment

 Take assessment online here: http://tinyurl.com/AnSci-HormoneFunction
Download and print the assessment by scanning this QR code or by going to this URL: http://www.tagmydoc.com/Ch17AnSci

18 Hormone Control

Major Concept

Hormones affect many vital functions in the body and are controlled by responses to the environment.

Objectives

- Define receptor cells, feedback, gland and homeostasis
- Identify pathways and hormones involved in the milk-ejection reflex
- Outline the way levels of hormones are controlled in the body

Key Terms

- Alveoli
- Epinephrine
- Feedback
- Homeostasis
- Hypothalamus
- Milk-ejection reflex
- Milk-letdown
- Oxytocin
- Posterior pituitary
- Receptor cells
- Thyroxine

Chapter Resource

Complementary *full color* illustrations, photos, charts and graphs are available by scanning this QR code or by following this URL: http://www.tagmydoc.com/AS18 This digital resource will enhance your understanding of the chapter concepts.

Hormones Control and Influence

- Hormones affect many vital functions in the body.

  o Amount of each hormone secreted varies with the state of the animal's body and the state of its environment (is it about to be milked, is it scared, etc.).

  o Since these items change constantly, the amount of hormone secreted must be carefully controlled so the animal can maintain normal **homeostasis** (balance of the body - temperature, heart rate, blood pressure, etc.).

- Endocrine glands that secrete hormones do so directly into the blood.

- Each hormone stimulates **receptor cells** (other cells that will respond to the hormone) in the target gland or organ.

- o Response of the target gland or organ depends on the level of the hormone in the blood and is carefully monitored by the body systems.

- **Milk-ejection reflex** is an example of endocrine gland activity.

  - o When the udder is stimulated, nerve impulses pass through the spinal cord to the brain.

  - o The brain responds by "telling" the **posterior pituitary** (back portion of the pituitary) to discharge **oxytocin** (a hormone released by the pituitary gland that causes increased contraction of the uterus during labor and stimulates the ejection of milk into the ducts of the mammary glands), into the bloodstream.

  - o Hormone is carried to the udder where it causes a muscle like contraction that brings about **milk-letdown** (release of milk into the teat cisterns).

  - o Milker or calf can easily obtain the milk that was made and held in the **alveoli** (milk producing cells).

  - o Stimulation of the udder occurs by washing by the milker, or suckling by the calf.

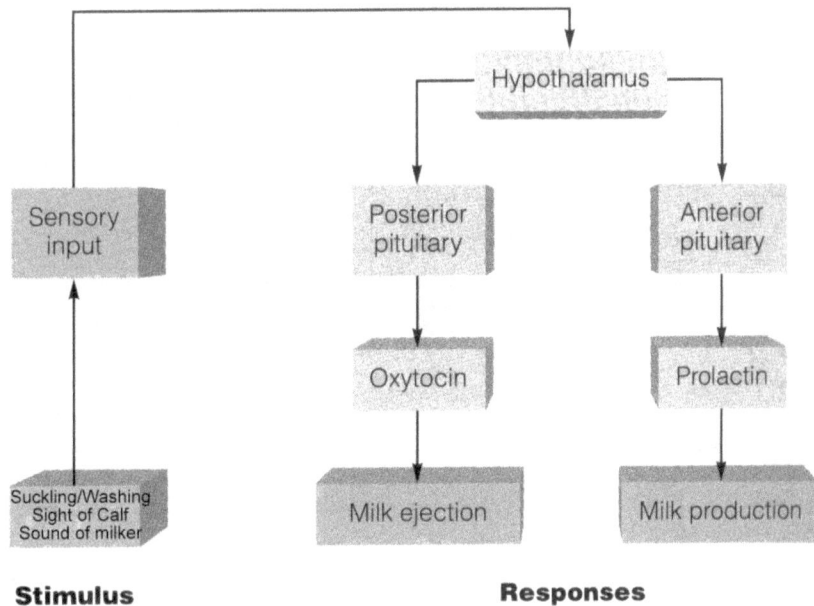

  - o If the cow is scared, or hurt (bitten by the calf or the milking machine, a nerve impulse will stimulate the adrenal gland to secrete **epinephrine** (adrenaline).

    - ✓ This hormone will overcome the action of oxytocin and shuts off the milk-ejection reflex.

- Other hormones

    o Needed at a constant level in the body are controlled by a continuous feedback system through the bloodstream and the brain through the hypothalamus.

    o Many hormones influence the growth, development and function of the mammary glands.

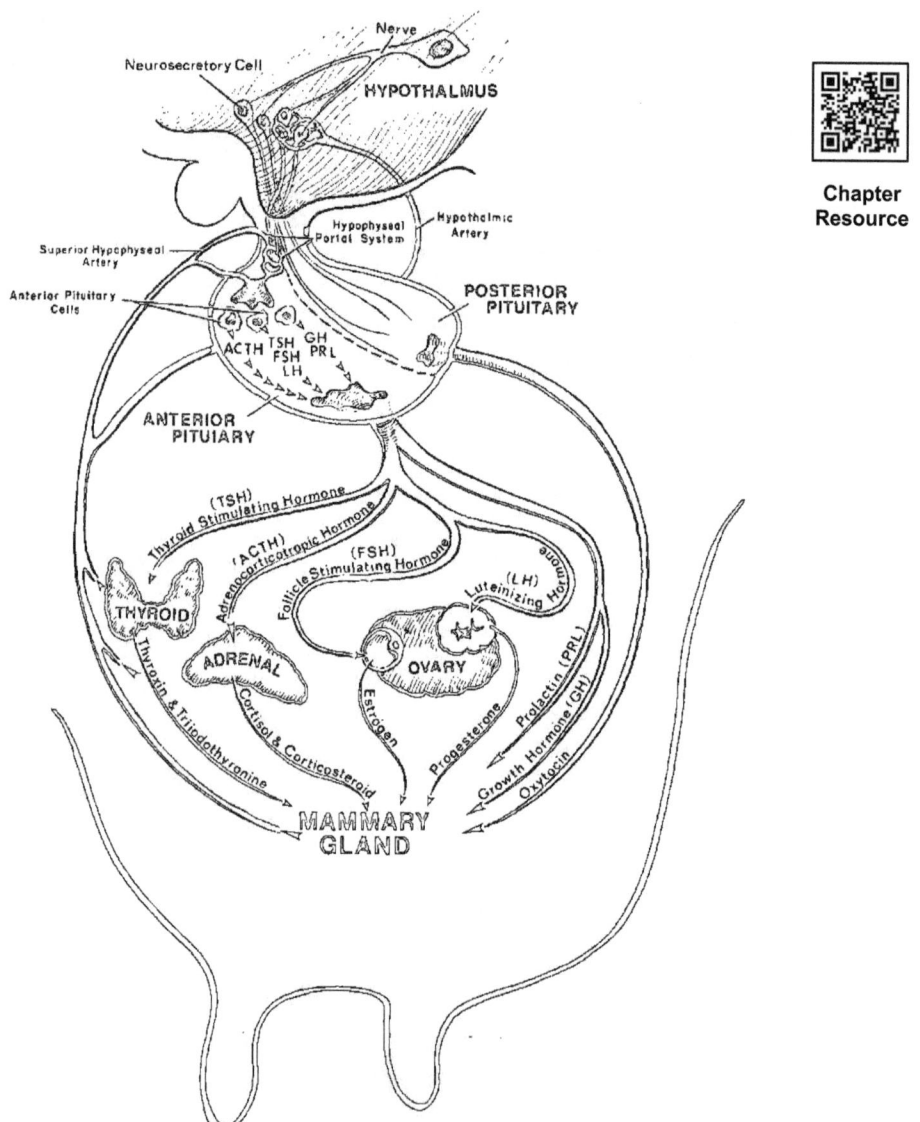

Chapter Resource

Feedback Control

- In this diagram, the level of the hormone thyroxine is controlled by negative feedback.

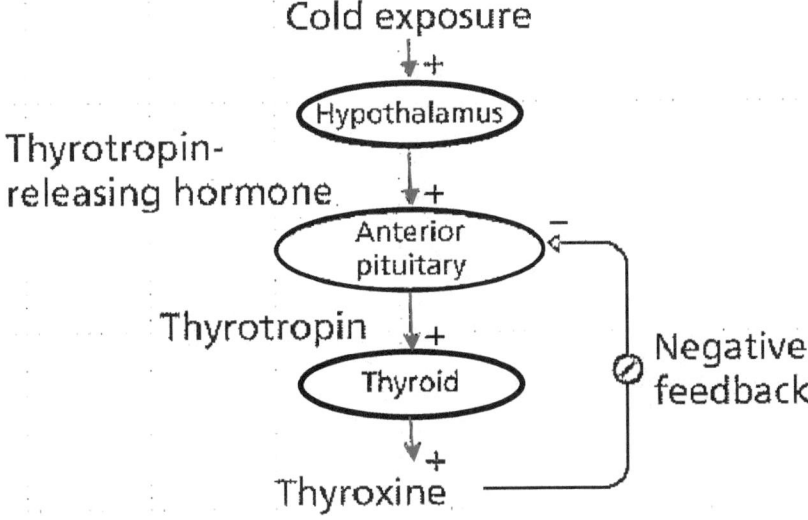

- The **hypothalamus** (a portion of the brain found in the floor of the third ventricle) regulates (controls) the level of thyroxin in the bloodstream in response to a stimulus (cold exposure).

- Receptor cells recognize thyroxine and monitor the level in the bloodstream all the time.

    - If the level is low, they "tell" the hypothalamus to secrete a releasing hormone that "tells" (stimulates) the pituitary to secrete thyroid stimulating hormone.

        - ✓ This puts the thyroid gland into action and releases thyroxin into the bloodstream.

    - When the level reaches normal again the receptor cells signal for shutdown of thyroxin release.

- Many hormones are controlled by feedback and in response to an animal's environment.

Summary

Hormones affect many vital functions in the body. The endocrine glands that secrete hormones do so directly into the blood. Each hormone stimulates receptor cells in the target gland or organ. Other hormones needed on a constant level in the body are controlled by a continuous feedback system through the bloodstream.

Additional Resources

Taylor, R.E. and T.G. Field. 2011. Scientific farm animal production. 10th ed. Upper Saddle River, NJ: Prentice Hall.

WikiBooks: Anatomy and Physiology of Animals -The Endocrine System
http://en.wikibooks.org/wiki/Anatomy_and_Physiology_of_Animals/Endocrine_System

WikiVet: Anatomy and Physiology - Endocrine System
http://en.wikivet.net/Category:Endocrine_System_-_Anatomy_%26_Physiology

Assessment

Take assessment online here: http://tinyurl.com/AnSci-HormoneControl
Download and print the assessment by scanning this QR code or by going to this URL: http://www.tagmydoc.com/Ch18AnSci

19 Reproductive Systems

Major Concept

Understanding the reproductive systems and processes of livestock is key to successful animal production.

Objectives

- Identify the major organs and their function in the reproductive tract of female and male livestock
- Name the four phases of the estrus cycle
- List the different roles of the female and male reproductive tracts

Key Terms

- Accessory sex glands
- Castration
- Cervix
- Cowper's gland
- Cryptorchid
- Diestrus
- Epididymis
- Estrus
- Fallopian tubes
- Lobules
- Metestrus
- Monorchid
- Motility
- Ovaries
- Proestrus
- Puberty
- Semen
- Seminal fluid
- Seminiferous tubules
- Sertoli cells
- Sigmoid flexure
- Spermatogenesis
- Spermatogonia
- Spermatozoa
- Testosterone
- Urethra
- Uterine horns
- Uterus
- Vagina
- Vas deferens
- Vasectomy
- Vulva

Chapter Resource

Complementary *full color* illustrations, photos, charts and graphs are available by scanning this QR code or by following this URL: http://www.tagmydoc.com/AS19 This digital resource will enhance your understanding of the chapter concepts.

Overview of Animal Reproduction

- Reproduction is a complicated process in all species of animals.

- Dependent upon the proper functioning of all physical and biochemical processes.

- Anatomy of both the male and female must be compatible.

- Physiological compatibility and timing is also required.

    o If two animals of different breeds or subspecies copulate, their systems (and genetic makeup) must be similar enough for the sperm and ovum to unite (fertilization) and begin cell division.

    o Female must be willing to accept the male (in heat).

    o Ovum must be mature and ready to be fertilized and ovulation must have occurred.

- Dependent on the proper function of many organs and under hormonal control.

- Any abnormality in the anatomy or physiology of the reproductive tract results in lower fertility or complete sterility of the animal.

Chapter Resource

Male Reproductive Tract

- Male's role is less complex than the female's.

    o Role is to produce large numbers of viable male sex cells called **spermatozoa**.

    o The male contributes half of the chromosomes to each of his offspring.

        ✓ After mating, the role of the male is over.

- Male reproductive organs in the various mammalian species are similar in form and function.

- Testes or testicles in the male are the primary sex organs and are held in a sac called the scrotum.

    o They produce spermatozoa and the male hormone testosterone.

    o Each male normally carries two testicles in the scrotum.

    o Scrotum functions as a heat regulating mechanism.

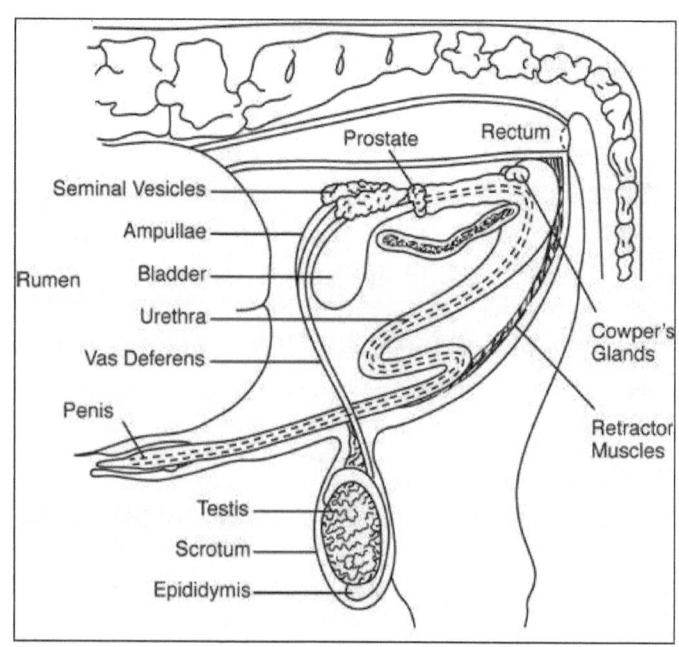

- Four to five degrees below body temperature is essential for **spermatogenesis** - the growth and maturation of the sperm.

• In some farm animals only one or neither testicle may drop out of the body cavity into the scrotum.

- A **monorchid** has one fertile testicle – one testicle dropped and the animal is fertile.
    - ✓ Although the testicle in the body does not produce viable spermatozoa, it does produce testosterone.
    - ✓ If both testicles remain in the body cavity the animal is termed a **cryptorchid** and is sterile.

• **Castration** of the male consists of removing both testicles.

Chapter Resource

- It is usually done early in life.
- Castrated males do not develop a sex drive and tend to gain weight quicker (for slaughter).
- When monorchids are castrated only the dropped testicle is taken and the animal retains its sex drive, but is infertile – an undesirable animal from a management standpoint.

• Each testicle is made of **lobules** (a small lobe) containing long tubes containing the spermatozoa in all stages of development.

- Sperm producing cells are called **spermatogonia**.
    - ✓ Cells surrounding the spermatogonia called **Sertoli cells** which serve a protective and nutritional role for the germ cells (spermatogonia or sex cells).
    - ✓ Sertoli cells are in microscopic tubes in the testes called **seminiferous tubules**.
- In the testes, the spermatozoa do not possess **motility** (ability to move under their own power) and are not capable of fertilization.
- **Epididymis** a long, convoluted tube on the backside of the mammalian testis where sperm mature and are stored; it consists of a head, body and tail which joins the vas deferens.
    - ✓ Head portion is connected to the testes.
    - ✓ Tail appears as an enlargement on the ventral portion of the testes in the scrotum.

- ✓ Tail is the storage for the spermatozoa (200 billion at a time).
- ✓ Spermatozoa mature as they migrate through the epididymis.
- ✓ The tail of the epididymis widens into a larger, longer tube, the vas deferens, leading to the urethra.

- **Vas deferens** leads from the tail of the epididymis into the urethra.
  - If a section of this is removed the male cannot reproduce.
  - This process is called a **vasectomy**.

- **Urethra** begins at the opening of the bladder and is continuous with the penis.
  - In mature bulls, the posterior portion of the urethra is S-shaped, known as the **sigmoid flexure**.
    - ✓ It extends the penis outside the body and into the vaginal cavity of the female so that the semen can be deposited.
    - ✓ If the sigmoid flexure does not work the animal is the equivalent of a sterile male.

- **Accessory sex glands** include:
  - Prostate gland
  - Two seminal vesicles
  - Two **Cowper's glands** (either of a pair of small glands that open into the urethra at the base of the penis and secrete a constituent of seminal fluid).
  - Function of the accessory sex glands is to provide an essential medium (fluid environment) called **seminal fluid** for the transport of the sperm from the testes to the vagina.
  - Seminal fluid functions:
    - ✓ Adds volume
    - ✓ Provides nutrients for the sperm
    - ✓ Cleans and flushes out the urinary tract
    - ✓ In some species (pigs and rats), makes a plug to hold the semen in the vagina.

Female Reproductive Tract

- Female plays a more important role in reproduction.

  o Provides half of the chromosomes of the young.

  o Nourishes the young in her **uterus** and after birth until weaning.

- Female tract consists of the vulva, vagina, cervix, uterus, uterine horns, fallopian tubes and ovaries.

  o **Vulva** is the exterior portion of the reproductive tract.

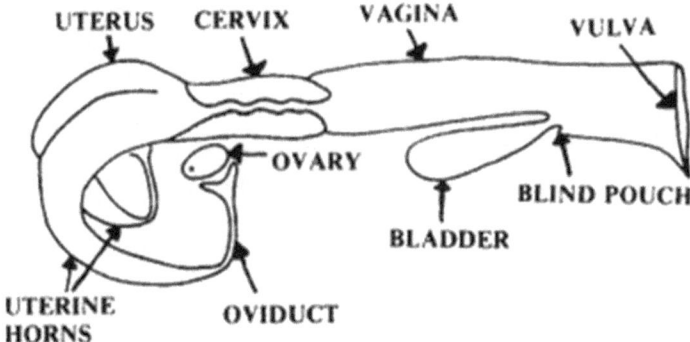

  o **Vagina** is the region between the vulva and the cervix.

    ✓ In cows and ewes, semen is deposited in the vagina.

    ✓ In mares and sows, the semen is deposited in the cervix.

  o **Cervix** is the opening into the uterus through which sperm must pass to fertilize the egg.

    ✓ At birth, the cervix stretches enough for the newborn to pass through.

    ✓ To help prevent entry of potentially harmful infections when pregnancy occurs, the cervix becomes blocked with a mucus plug.

  o **Uterine horns** are the two branches of the uterus.

  o Forward end of each horn leads to a small convoluted tube, called the Fallopian tube or oviducts.

  o **Fallopian tubes** or oviducts are found at the anterior end of each uterine horn.

    ✓ Lined with microscopic cilia that assist the ovum (egg) as it travels down toward the uterine horn.

- ✓ The normal female has two **ovaries** located in close proximity to each fallopian tube.
  - o Each ovary possesses many female reproductive cells (ova) in all stages of development.
    - ✓ Unlike the male who produces all sperm after the onset of **puberty** (sexual maturity), each female possesses all her potential ova at the time of birth.
    - ✓ Relatively few reach maturity and are ovulated.
    - ✓ If fertilization does not occur after ovulation, the ovum is gradually reabsorbed.
    - ✓ Ovulation is under hormonal control.

Reproduction in Males

- After sexual maturity of the male, sperm production is continuous under most conditions.
  - o **Testosterone** is the primary male sex hormone responsible for:
    - ✓ Growth, development and secretory activity of the accessory glands
    - ✓ Survival of spermatozoa
    - ✓ Secondary sex characteristics, crest, male voice and sex drive
  - o **Semen** is the term used for the fluid ejaculated by the male for reproductive purposes. Made up of:
    - ✓ Fluid from the many accessory glands
    - ✓ Sperm cells

    Chapter Resource

  - o Number of spermatozoa and total volume of seminal fluid varies in each species.
  - o Spermatozoa may be stored in the epididymis for several days.
    - ✓ In the female, sperm will survive for 24-96 hours.
    - ✓ If frozen for artificial insemination, it will last many years.

Reproduction in Females

- Most farm animals reach sexual maturity in 4 (sow) to 24 (mare) months; at this time, the follicles begin to develop.

- Female comes in heat (estrus).
- Follicle containing the maturing ovum bursts.
- Egg is ovulated and travels down the oviducts to a point at which it can be fertilized by the sperm.

- The age of puberty not only varies between species, but also among individuals within the same species.
  - Puberty is brought on by hormone secretions, primarily of estrogen.

Estrus and the Estrous Cycle

- Estrus in farm animals is the time the female has or is about to ovulate and is receptive to the male.

- First estrous cycle begins at puberty and is the interval between two estrous periods while the animal is not pregnant. (Note: It is spelled estrous for the cycle and estrus for the time of receptivity.)

- Estrous cycle phases:

  - **Proestrus**, the phase just before estrus:
    - ✓ Vaginal wall thickens in preparation to receive the fertilized egg.
    - ✓ Ovary is about to release the ovum.

  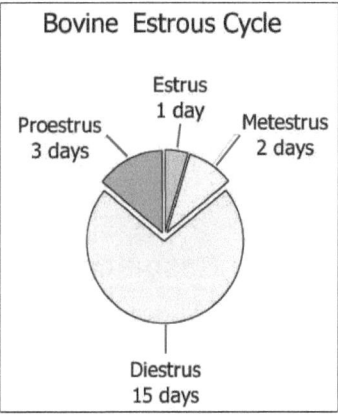

  - **Estrus** is the time for acceptance of the male and normally coincides approximately with ovulation.

  - **Metestrus,** the period immediately following estrus.
    - ✓ If conception has taken place the uterus prepares for pregnancy.
    - ✓ Fertilized egg will settle and attach there.

  - **Diestrus** occurs between metestrus and proestrus and is the longest period of the cycle.

- Non-pregnant females normally express both inward and outward signs of heat throughout the year:
  - Noise and aggressive behavior.

- o Swelling of the external portions of the reproductive tract.
- o Color change of the vaginal wall.
- Estrous cycles
  - o Stop after conception and begin again soon after parturition (birth).
    - ✓ Sows, 1 to 5 days post-partum
    - ✓ Mares, 5 to 10 days post-partum
    - ✓ Cows, 30 to 60 days post-partum
    - ✓ Ewes are seasonal breeders

Chapter Resource

Summary

Reproduction is a complicated process in all species of animals and is dependent upon the proper functioning of all physical and biochemical processes of the animal. The anatomy of both the male and female must be compatible. Any abnormality in the anatomy or physiology of the reproductive tract results in lower fertility or complete sterility of the animal. The male's role is to produce large numbers of viable male sex cells called spermatozoa. The male contributes half of the chromosomes to each of his offspring. The testes or testicles in the male are the primary sex organs and are held in a sac called the scrotum. Castration of the male consists of removing both testicles. The female tract consists of the vulva, vagina, cervix, uterus, uterine horns, fallopian tubes and ovaries. Most farm animals reach sexual maturity in 4 (sow) to 24 (mare) months. The first estrous cycle begins at puberty and is the interval between two estrous periods while the animal is not pregnant.

Additional Resources

Taylor, R.E. and T.G. Field. 2011. Scientific farm animal production. 10th ed. Upper Saddle River, NJ: Prentice Hall.

The Drost Project: Visual Guides to Reproduction
http://drostproject.org/

WikiBooks: Anatomy and Physiology of Animals - Reproductive System
http://tinyurl.com/lha9eah

Assessment

 Take assessment online here: http://tinyurl.com/AnSci-ReproductiveSystem
Download and print the assessment by scanning this QR code or by going to this URL: http://www.tagmydoc.com/Ch19AnSci

20 Fertilization

Major Concept

Fertilization is the first step in successful livestock production.

Objectives

- Outline the process of gametogenesis in the male and female
- Define the haploid, diploid, segregation and zygote
- List the steps in fertilization

Key Terms

- Capacitation
- Diploid
- Gametogenesis
- Haploid
- Oocyte
- Oogenesis
- Polar bodies
- Pronuclei
- Spermatogenesis
- Spermatogonia
- Zona pellucida
- Zygote

Chapter Resource

 Complementary *full color* illustrations, photos, charts and graphs are available by scanning this QR code or by following this URL: http://www.tagmydoc.com/AS20 This digital resource will enhance your understanding of the chapter concepts.

Gametogenesis

- Formation gametes

- Sperm in males

- Egg in females

- Process of cell division called meiosis

- Gametes contain one half of the genetic material found in normal body cells called "1n" or **haploid**.

Spermatogenesis

- Process by which male sex cells are produced.

- Occurs continuously once the animal reaches puberty.

- Once a cell begins spermatogenesis it takes approximately 60 days to turn into a mature sperm cell ready to fertilize an egg.

- Sperm are formed within the seminiferous tubules from cells called spermatogonia.

- Process is complex and involves cell division and differentiation.

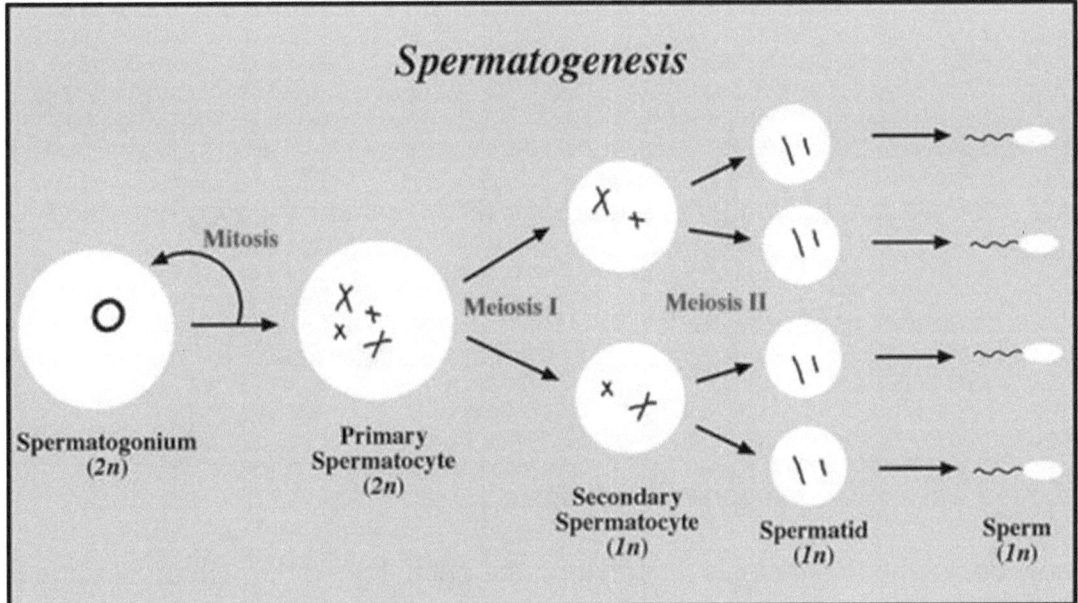

- Final cell has half the number of chromosomes of a normal "somatic" or body cell, making it a sex or "germ" cell.

- Under hormonal control

Oogenesis

- Process by which the female egg is produced.

- In the female, all eggs are present in the ovaries by the time the animal is born.

- Egg at this point is called an **oocyte** and is in a resting stage after completing meiosis I.

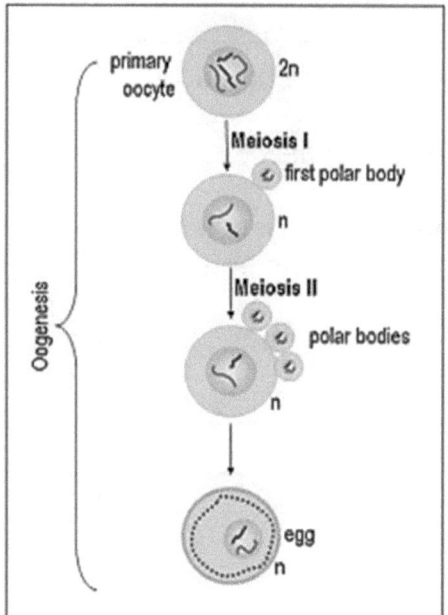

- Once a female reaches puberty, a hormone (FSH) causes an oocyte to begin meiosis.

- Process is under hormonal control.

Oocyte

- Ovulated while in the metaphase of meiosis II.

- A new oocyte begins meiosis II each time the animal goes through an estrous (heat) cycle.

- A sperm fertilizing the oocyte causes the completion of meiosis II.

- Remaining genetic material cast off as **polar bodies** (minute cell produced and ultimately discarded in the development of an oocyte).

Chapter Resource

Fertilization

- Fusion of male and female gametes to form a single cell, the **zygote**.

Female Gamete: The Ovum

- The ova are in metaphase of Meiosis II at the time of ovulation.

- Surrounded by the **zona pellucida** which is a protective covering around the ova.

- Usually fertile for about 25 hours.
- Travels to the upper third of the oviduct where it is fertilized, if sperm are present.

Male Gamete: The Sperm

- Millions of sperm are found in each ejaculate, but only a few thousand will reach the site of fertilization.

- Only one sperm can penetrate and fertilize the ova.

- Sperm can reach the site of fertilization as quickly as within 15 minutes of ejaculation.

- In some species, spermatozoa undergo changes in the female tract before fertilization.

    o This process is called **capacitation**; this activates the sperm so it can fertilize the egg. This occurs in rabbits, mice and sheep.

- Fertile life of a sperm is 2-3 days.

Chapter Resource

Fertilization Process

- Entry of sperm into ovum.

    o Sperm must penetrate the zona pellucida of the ovum. This causes the completion of meiosis II in the ovum.

    o Sperm head attaches to the membrane that surrounds the ova and the genetic material from the sperm is taken into the cell (which already is holding the female's contribution of genetic material).

- Pronuclei Formation

    o Male and female **pronuclei** (either of a pair of gametic nuclei, in the stage following meiosis but before their fusion leads to the formation of the nucleus of the zygote) are seen as separate entities before uniting to form a single nucleus. Then the cell again has pairs of chromosomes or the condition called "2n" or diploid.

    o Once the pronuclei have united, the process of fertilization is complete.

    o Zygote now begins to undergo mitotic divisions to develop into the embryo.

Visual Representation of Gametogenesis, Chromosome Count and Fertilization

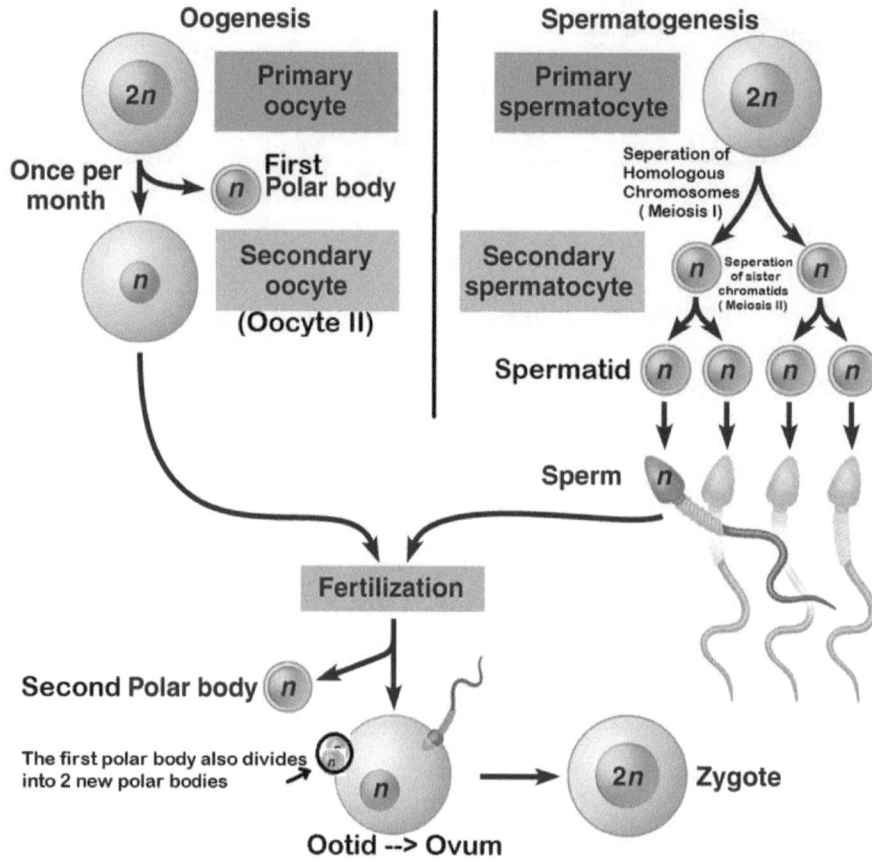

Segregation

- Separation, during **gametogenesis** of the pairs of chromosomes. If there are different "instructions" on each of the two homologous (similar in structure, controlling the same traits) one set of instructions goes to one gamete and the other set carried on the other chromosome goes to the other gamete.

- Segregation or separation begins during the reduction division shown below, and is completed when each germ cell has been produced and is in a haploid (1n) state.

Summary

Spermatogenesis is the process by which male sex cells are produced. Sperm are formed within the seminiferous tubules from cells called spermatogonia. Once a female reaches puberty, a hormone causes an oocyte to begin meiosis, a process called oogenesis. A new oocyte begins meiosis II each time the animal goes through a heat cycle. Fertilization is the fusion of male and female gametes to form a single cell, the zygote. The ova is in metaphase of meiosis II at the time of ovulation. Millions of sperm are found in each ejaculate, but only a few thousand will reach the site of fertilization. Segregation is the separation, during gametogenesis of the pairs of chromosomes. If there are different "instructions" on each of the two homologous (similar in structure, controlling the same traits) one set of instructions goes to one gamete and the other set carried on the other chromosome goes to the other gamete.

Additional Resources

Taylor, R.E. and T.G. Field. 2011. Scientific farm animal production. 10th ed. Upper Saddle River, NJ: Prentice Hall.

Fertilization in Animals Reproduction in Animals Part II
www.youtube.com/watch?v=sWYZPJ4UXHg

Assessment

Take assessment online here: http://tinyurl.com/AnSci-Fertilization
Download and print the assessment by scanning this QR code or by going to this URL: http://www.tagmydoc.com/Ch20AnSci

21 Conception, Gestation and Parturition

Major Concept

Normal processes of conception, gestation and parturition are critical for successful livestock production.

Objectives

- Outline the process of fertilization
- Provide the gestation time for three animals
- Identify three signs of impending parturition (birth)
- List two concerns during parturition
- Identify the role of the placenta

Key Terms

- Afterbirth
- Capacitation
- Colostrum
- Distended
- Endometrium
- Fertilization
- Fetus
- Gametes
- Gestation
- Implantation
- Ova
- Ovaries
- Oviduct
- Parturition
- Placenta
- Zygote

Chapter Resource

Complementary *full color* illustrations, photos, charts and graphs are available by scanning this QR code or by following this URL: http://www.tagmydoc.com/AS21 This digital resource will enhance your understanding of the chapter concepts.

Conception, Sex Cells and Fertilization

- Sex cells are collectively called **gametes**. They contain half of the number of chromosomes (haploid) as the other body cells, making them unique in the body.

    o Sex cells are produced by meiosis in the **ovaries** (a female reproductive organ in which ova or eggs are produced) or testes.

    o In the male testes, meiosis produces sperm cells.

- In the female ovaries, meiosis produces egg cells or **ovum** (a mature female reproductive cell).

- **Fertilization** is the action or process of fertilizing an egg, female animal, or plant, involving the fusion of male and female gametes to form a **zygote**.

    - Occurs in the upper third of the fallopian tube or **oviduct** in the female.
    - Sperm may reach the ovum 15-20 minutes after mating.
    - Sperm travels 3-4 mm per minute; their progress is aided by uterine contractions.
    - In some species, spermatozoa undergo changes in the female tract before fertilization.
        - ✓ This is called **capacitation** and activates the sperm so it can fertilize the egg.
    - Although millions of spermatozoa are introduced, only a few reach the site of fertilization.
    - Ordinarily, only one sperm can fertilize an egg.
    - Transport of the fertilized ovum through the oviduct takes three to five days to reach the uterus.
    - Once fertilization (the union) has occurred, the chromosomes in the newly formed nucleus control the cell activities and the eventual form, size, etc. of the individual.

- **Implantation** occurs when the zygote embeds into the **endometrium** (mucous membrane that lines the uterus).

    - Interval between arrival in the uterus and implantation varies.
        - ✓ Three days in rodents
        - ✓ About seven weeks in a mare
        - ✓ Three to four weeks in the cow
        - ✓ Eleven to 14 days in the ewe
        - ✓ Four to seven days in the sow
        - ✓ Three days in women

    Chapter Resource

    - At the time of implantation, the fetal (extraembryonic) membranes form the placenta.

Placenta

- Structure attached to the inner surface of the uterus through which the embryo or **fetus** obtains its nourishment and discharges its wastes.

    o Produces hormones – estrogen and progesterone

- Type varies by species:

    o Cow, ewe, doe (goat): cotyledonary

    o Mare, sow: diffuse

    o Doe (rabbit): discoid

    o Bitch (dog): zonary

- Expelled at birth and called **afterbirth**.

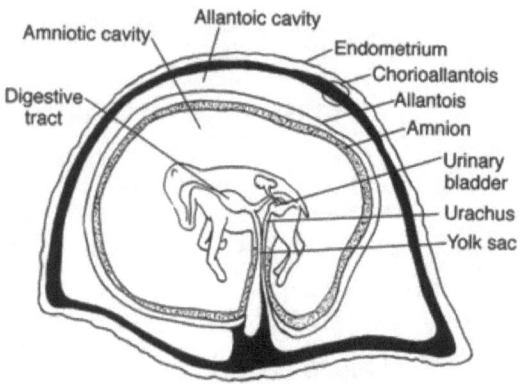

Equine fetus in the placenta

Gestation

- Time from conception to birth.

- Varies for different livestock species.

    o Cow – 285 days

    o Mare – 330 days

    o Sow – 114 days

    o Ewe – 148 days

    o Doe (goat) – 150 days

- Number of young varies by species

    o Cow – usually one

    o Mare – usually one

    o Sow – litters of up to 10-15

    o Ewe – usually one but can be two or three

    o Doe (goat) – one or two

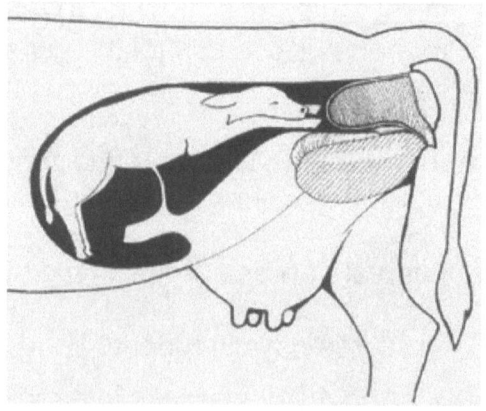

Parturition

- The act or process of giving birth.

    o Initiated by the release of the hormone cortisol from the fetal adrenal cortex.

    o Progesterone levels decline.

    o Estrogen, prostaglandin and oxytocin levels rise.

    o Uterine contractions increase.

    o Cervix relaxes.

    o Ligaments and cartilage in pelvic region soften and relax due to hormone relaxing.

Chapter Resource

- Signs of Approaching Parturition:

    o **Distended** (swollen) udder.

    o Female becomes extremely nervous and uneasy, often will separate from herd.

    o As parturition approaches fluid contents of udder change from a watery solution to the thick, milky **colostrum**.

    o Noticeable enlargement and swelling of vulva.

    o Muscular contractions are evident as labor begins.

    o First the water bag will appear on the outside, increasing in size until it ruptures.

- Most common presentation is feet first, followed by nose, shoulders, middle, hips, hind legs and feet for cows, sheep and horses. Piglets do not orient themselves in any one direction.

- A vigorous newborn will stand in about 15 minutes and be nursing within about 30 to 60 minutes – quicker for piglets.

- Three to six hours after parturition the placenta, or afterbirth, is expelled.

    o If it remains in for more than 24 hours, manual removal should be done by an experienced caretaker or veterinarian/technician.

- Concerns
    - Labor without progress
    - Posterior presentation
    - Retaining of placenta

Summary

Fertilization consists essentially of the fusion of two cells, the male and female gametes, to form one single cell, the zygote. Implantation occurs when the zygote is embedded into the endometrium of the uterus and the placenta forms. Gestation is the time from fertilization to birth and varies with livestock species. Parturition is the act or process of giving birth. Normal presentation is front feet first, followed by nose which is resting on them, shoulders, middle, hips, hind legs and feet - except for piglets. A vigorous newborn will attempt to stand and nurse within minutes. Shortly after parturition the placenta is expelled.

Additional Resources

Parker, R. 2013. Equine science. 4th ed. Clifton Park, NY: Delmar Cengage Learning.

Management of Dairy Heifers – Penn State
http://extension.psu.edu/animals/dairy/nutrition/heifers/heifer-feeding-and-management/management-of-dairy-heifers

Physiology of Pregnancy
http://www1.agric.gov.ab.ca/$department/deptdocs.nsf/all/beef4892

Assessment

Take assessment online here: http://tinyurl.com/AnSci-Conception
Download and print the assessment by scanning this QR code or by going to this URL: http://www.tagmydoc.com/Ch21AnSci

22 Inheritance

Major Concept

Nucleic acids, DNA and RNA, play a role in inheritance.

Objectives

- List the principles of genetic inheritance discovered by the work of Gregor Mendel
- Identify the relationships between chromosomes and DNA, and DNA and genes
- List the chemical compounds that make up DNA
- Name the four nitrogenous bases in DNA
- Recognize the role of DNA in protein synthesis

Key Terms

- Chromosomes
- Codons
- Diploid
- DNA
- Gene
- Genetic
- Haploid
- Nitrogenous
- Nucleotide
- RNA
- Sequencing

Chapter Resource

Complementary *full color* illustrations, photos, charts and graphs are available by scanning this QR code or by following this URL:
http://www.tagmydoc.com/AS22 This digital resource will enhance your understanding of the chapter concepts.

Brief History of Genetics

- 1670's:

  o Scientists believed that each sperm contained a "little man" that would develop into a human. The mother only served as an incubator.

- 1750's:

  o Scientists believed in the Blending of Inheritance Theory – For example, a black animal mated to a white animal would produce a gray animal.

- 1850's: Gregor Mendel

- An Austrian monk, who did a series of experiments that led to a new understanding of the mechanism of inheritance.

- Worked in the monastery garden mating pea plants.

- Determined that characteristics were inherited by discrete factors that would eventually become known as genes.

Chapter Resource

- Through his work, Mendel discovered (described) the following principles:

  - Principle of Segregation: Every individual carries pairs of factors for each trait and the members of the genes segregate at random (without specific control or predetermined plan) during the formation of gametes.

    - ✓ Since segregation is random, predictable ratios of traits are found in the offspring.
    - ✓ During segregation chromosomes go from a pair (2n or diploid state) to singles (1n or haploid state).
    - ✓ A good way to remember the difference: "hap a diploid is 1n".

  - Genes do not blend and are passed intact from one generation to the next and chance governs the segregation of the chromosomes in the gametes.

  - Principle of Independent Assortment: Members of each pair of genes are distributed independently when the gametes are formed and are unaffected by other gene pairs on other chromosomes.

    - ✓ For example: When pairs of genes on different chromosomes separate, they have an equal chance or probability of going to an individual gamete. There is no predetermined order for the dividing pairs – it's every gene for itself. (Note: This is the reason for genetic variation).

Chromosomes and DNA

- **Chromosomes** are in the nucleus of the cell and contain all the **genetic** (of or relating to genes or heredity) material in the cell.

- Chromosomes are arranged in pairs.

  - **Haploid**: Half the diploid or somatic (non-sex cell or gamete) number of chromosomes (n or 1n).

  - **Diploid**: Number of chromosomes found in the somatic or body cells (2n). Also, can be considered as twice the number of chromosomes found in the gametes (sperm or ovum).

Species	n	2n
Cattle	30	60
Sheep	27	54
Goat	30	60
Swine	40	80
Horse	30	60
Human	23	46

- Chromosomes are made up of a substance called "DNA."

DNA: Deoxyribonucleic Acid

- DNA is composed of three components;

    1. Deoxyribose sugar
    2. Phosphate
    3. Four nitrogenous bases (adenine, thymine, guanine and cytosine)

- A **nucleotide** is the combination of the deoxyribose, phosphate and one of the four bases.

    o Nucleotides bond together to form one strand of the DNA molecule.

    o Two of these strands wind around each other in a double helix (like a twisted ladder) to form the DNA molecule.

    o In 1953, Watson & Crick published findings based on X-ray analysis and other data that DNA was in the form of a "Double Helix".

- Bases of the DNA hold the key to inheritance. The four **nitrogenous** (means they contain the element nitrogen) bases are:

    1. Adenine (A)
    2. Thymine (T)
    3. Guanine (G)
    4. Cytosine (C)

- In the two strands of DNA, A is <u>always</u> paired with T, and G is <u>always</u> paired with C.

    o Example: TAAGGAAGTACTTACAT

    ATTCCTTCATGAATGTA;

- Replication of DNA – (How it copies itself during mitosis or meiosis.)

    o Double helix unwinds and pulls apart.

- A new strand is formed using the old strand as a template.
- Result is two identical double helix strands; in three steps:
  1. One DNA helix
  2. The helix begins to split
  3. New bases come in and attach as the helix unwinds
- Two identical DNA molecules originating from the one.

A-T	A-T
A-T	A-T
T-A	T-A
G-C	G-C
C-G	C-G
G-C	G-C
T-A	T-A
A-T	A-T

Chapter Resource

Genes

- Definition: Genes are points of activity found in each chromosome that govern the way in which traits develop.
- Genes are specific areas on each chromosome and are made up of DNA.

Genetic Codes and Protein Synthesis

- Protein synthesis in cells of the body guides the anatomy and the physiology of an animal.
- Genetic information is stored in the DNA molecule in the form of **codons.**
  - A codon is a triplet (3) of nucleotides bases.
  - For example, the DNA triplet CGC codes the amino acid, alanine. (Remember that protein molecules are built of individual amino acids).
- Each different possible code represents a protein.
- **RNA** (ribonucleic acid) is a group of molecules in charge of "reading" and "translating" the genetic code for the formation of new proteins.

- o RNA uses the DNA as a template to read the code in order to produce the right protein with the correct order and number of amino acids.

- Three kinds of RNA

    1. Transfer RNA (tRNA) plays a key role in protein synthesis (building).

        ✓ Each tRNA molecule can combine with one amino acid and can transport the a.a. to the new protein building site in the cytoplasm of the cell.

    2. Ribosomal RNA (rRNA) also plays a key role in protein synthesis.

        ✓ It helps control the connecting of the parts of the protein (the amino acids) together.

    3. Messenger RNA (mRNA) which helps complete the building of the protein.

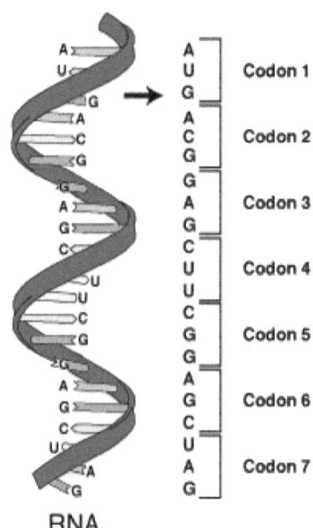

RNA
Ribonucleic acid

   ✓ Does so by physically **sequencing** (putting the amino acids in correct order) the amino acids that were carried to the building site by the tRNA and chemically connected by the rRNA.

   ✓ mRNA directs the sequence based on the order it obtains from the DNA molecules.

- Genetic code and protein synthesis can be compared to the building of a house.

    o tRNAs are the people who haul the building materials to the construction site in the suburbs (cytoplasm – on the ribosomes).

    o rRNAs are people that drive the nails, pour the concrete, etc.

    o mRNAs provide a physical pattern or blueprint (that is a copy of the original) for the construction of the house.

    o DNA is the original blueprint that is kept "downtown" in the main office (in the nucleus) that directed the plan that is physically duplicated by the mRNA.

Summary

Mendel was an Austrian monk, who did a series of experiments that would lead to a new understanding of the mechanism of inheritance, including the Principle of

Segregation and the Principle of Independent Assortment. Chromosomes are located in the nucleus of the cell and contain all of the genetic material in the cell. DNA is composed of three components: deoxyribose sugar; phosphate; and four nitrogenous bases. The bases of the DNA hold the key to inheritance. The four nitrogenous bases are: adenine (A), thymine (T), guanine (G) and cytosine (C). Genes are specific areas on each chromosome and are made up of DNA. Genetic information is stored in the DNA molecule in the form of codons.

Additional Resources

Parker, R. 2013. Equine science. 4th ed. Clifton Park, NY: Delmar Cengage Learning. pp. 217-227.

DNA and Protein Synthesis on YouTube
http://www.youtube.com/playlist?list=PLBD0AFAD964A33D6E

From DNA to Protein
https://www.youtube.com/watch?v=gG7uCskUOrA

Khan Academy – Chromosomes, chromatids, chromatin, etc.
http://tinyurl.com/c44zfaw

Assessment

Take assessment online here: http://tinyurl.com/AnSci-Inheritance
Download and print the assessment by scanning this QR code or by going to this URL: http://www.tagmydoc.com/Ch22AnSci

23 Interaction of Dominant and Recessive Genes

Major Concept

An understanding of genes and probability allow charting of the possible results of a mating involving dominant and recessive genes.

Objectives

- Determine genotypic and phenotypic ratios in simple matings involving dominant and recessive genes
- Set up a Punnett Square
- Identify alleles each parent can contribute in simple cases of dominance and recessive
- Define incomplete dominance and sex-linked inheritance

Key Terms

- Allele
- Diploid
- Gametogenesis
- Haploid
- Heterozygous genotype
- Homozygous
- Sex-linked inheritance
- Incomplete dominance
- Dominate
- Recessive

Chapter Resource

Complementary *full color* illustrations, photos, charts and graphs are available by scanning this QR code or by following this URL: http://www.tagmydoc.com/AS23 This digital resource will enhance your understanding of the chapter concepts.

Characteristics of Genes

- Traits are controlled by genes.

    o Genes have different forms which are known as **alleles**.

    o For example, the gene for coat color could have two alternatives – a white allele and a black allele.

- o In **diploid** (containing two complete sets of chromosomes, one from each parent) organisms, such as man and higher animals, there are two alleles – one on each chromosome of the homologous pair.

- o One of each of the above example alleles (black/white) could be found on a pair of homologous chromosomes.

- o When those chromosomes undergo segregation during **gametogenesis** (process in which cells undergo meiosis to form gametes), one gamete would obtain the allele for white and the other the allele for black. Thus, the resulting germ cells could pass on either the allele for black or the allele for white, but not both because of segregation of the homologous chromosomes.

- **Dominant Genes**

  - o A dominant gene covers (masks) the presence of other genes.

  - o Capital letters are used to designate dominant genes.

  - o Using the black/white coat color example – black could be "dominant" over white.

- **Recessive Genes**

  - o A recessive gene is the one over-ridden by the dominant gene.

  - o For a recessive trait to appear (phenotypically) the animal must be **homozygous** (have two of the same gene – one on each homologous chromosome) for that characteristic to appear.

  - o Recessive genes are designated by small letters.

  - o In the example in "B" above, white is "recessive" to black.

Chapter Resource

- Incomplete Dominance

  - o Sometimes traits do not have one clear dominant gene or one clear recessive gene.

  - o In incomplete dominance, traits appear to blend together.

  - o When crossing a red snapdragon with a white snapdragon, the offspring is pink if **incomplete dominance** (a form of intermediate inheritance in which one allele for a specific trait is not completely dominant over the other allele) occurs.

  - o **Heterozygous genotype** (an organism that has both the dominant and the recessive gene) incomplete dominance appears different than either of the homozygous alternatives. (For example, crossing plants with red or white flowers produces plants with pink flowers.)

One Pair of Genes

- Punnett squares can provide phenotype, genotype, phenotypic ratio and genotypic ratio. For example:

  o Angus cattle; coat color

      ✓ Black (B) is dominant
      ✓ Red (b) is recessive

Chapter Resource

  o Cross a homozygous black cow to a homozygous red bull.

      ✓ BB x bb.
      ✓ The cow can only give "B" alleles (for black coat color) to the offspring and bull can only give "b" alleles (for red coat color) to the offspring.
      ✓ Set up Punnett Square.

Cow/bull gametes	B	B
b	Bb	Bb
b	Bb	Bb

      ✓ All offspring genotype is Bb with a black phenotype only.

  o Cross a Bb cow to a Bb bull. In this case, each parent provides gametes with "b" or "B". There are two alternatives.

      ✓ Determine what alleles each parent can donate and in what ratio.
      ✓ Set up Punnett square. Gametes being **haploid** (1n), around the outside of the square and the resulting combinations, the zygotes, being diploid (2n), in the cells in the Punnett Square.

Cow/bull gametes	B	b
B	BB	Bb
b	Bb	bb

      ✓ Determine genotypic ratio
      ✓ 1/4 BB: 2/4 Bb: 1/4 bb   1:2:1 ratio
      ✓ (25%): (50%) (25%)
      ✓ Determine phenotypic ratio

- ✓ 1/4 Bb + 1/4 Bb + 1/4 BB = 3/4 black + 1/4 bb= red
- ✓ 3:1 ratio

Two Pairs of Genes

- Angus Cattle: coat color and polled/horned.

  - Coat color: B= black; b= red

  - Horns: P= polled; p= horned

  - A Black Angus bull is heterozygous for the dominant gene for black coat (Bb) and heterozygous for the dominant polled gene (Pp). This bull (BbPp) is bred to Red Angus cows (bbPp) homozygous for the recessive coat color (bb) heterozygous for dominant polled gene (Pp).

  - Describe all the possible genotypes and phenotypes from this mating.

  - Use a Punnett Square to predict the possibilities:

Possible gametes from the cows (bbPp)	Possible gametes from the bull (BbPp)			
	BP	Bp	bP	bp
bP	BbPP	BbPp	bbPP	bbPp
bp	BbPp	Bbpp	bbPp	bbpp
bP	BbPP	BbPp	bbPP	bbPp
bp	BbPp	Bbpp	bbPp	bbpp

  - Phenotypes

    - ✓ Six calves will appear as Red Angus polled.
    - ✓ Two calves will appear as Red Angus horned.
    - ✓ Six calves will appear as Black Angus polled.
    - ✓ Two calves will appear as Black Angus horned.

  - Genotypes and number of each:

Genotype	Number
BbPP	2
BbPp	4
Bbpp	2
bbPP	2
bbPp	4
bbpp	2

Sex Linked Inheritance

- Phenotypic expression of an allele related to the chromosomal sex of the individual.

- In mammals, the female is the homogametic sex, with two X chromosomes (XX), while the male is heterogametic, with one X and one Y chromosome (XY).

- Genes on the X or Y chromosome are called sex-linked. For example:

  o Fur color in domestic cats.

    ✓ Gene that causes orange pigment is on the X chromosome.
    ✓ Calico or tortoiseshell cat, with both black (or gray) and orange pigment, is nearly always female.

  o Color blindness in humans.

Heredity vs Environment

- All traits of livestock are not controlled by single gene pairs.

- Most economically important traits are controlled by many pairs of genes.

- These traits are, in turn, influenced by the environment.

- Traits controlled by multiple genes range in influence due to genetics from 10% to 50%; meaning environmental conditions can affect traits from 90-50%.

Summary

In heterozygous situations, when those chromosomes undergo segregation during gametogenesis, one gamete obtains the dominant allele and the other the recessive allele for a trait. Thus, the resulting germ cells could pass on either the dominant allele or the recessive allele, but not both because of segregation of the homologous chromosomes. Punnett Squares are useful in predicting the genotypic and phenotypic ratios in simple cases of dominance and recessive. All traits of livestock are not controlled by single gene pairs.

Additional Resources

Parker, R. 2013. Equine science. 4th ed. Clifton Park, NY: Delmar Cengage Learning. (Pg. 217-234).

Epistasis: Gene Interaction and Phenotype Effects
http://www.nature.com/scitable/topicpage/epistasis-gene-interaction-and-phenotype-effects-460

Genetic Traits in Cattle
http://oklahoma4h.okstate.edu/aitc/lessons/intermed/hairy.pdf

Assessment

Take assessment online here: http://tinyurl.com/AnSci-Interaction
Download and print the assessment by scanning this QR code or by going to this URL: http://www.tagmydoc.com/Ch23AnSci

24 Understanding Genotype and Phenotype

Major Concept

All animals and plants have an identifiable genotype and phenotype that is passed on to their next generation.

Objectives

- Define the terms phenotype, genotype, homozygous, heterozygous, dominant, recessive and allele
- Construct and analyze a simple Punnett square

Key Terms

- Allele
- Dominant
- Gene
- Genotype
- Heterozygous
- Homozygous
- Phenotype
- Recessive

Chapter Resource

Complementary *full color* illustrations, photos, charts and graphs are available by scanning this QR code or by following this URL: http://www.tagmydoc.com/AS24 This digital resource will enhance your understanding of the chapter concepts.

Introduction

- Gregor Mendel's (1822-1884) experiments with pea plants began the understanding of genetic inheritance and the concepts of phenotype and genotype.

- Terms to understand:

  o **Allele**: A pair of genes which are located at the same place on homologous chromosomes (the matched pair); for example: in a chromosome pair, one chromosome has the gene or allele for red flowers and the other has the allele for white flowers. Also, is often applied to the traits associated with the genes.

  o **Phenotype**: The physical appearance of the animal because of its genetic makeup (genotype).

- **Gene**: A specific region of a chromosome which can determine the development of a specific trait that is composed partially or wholly of DNA; for example, red or white flowers.

- **Genotype**: Genetic constitution (makeup) of an individual.

- **Dominant**: Refers to genes which hide or mask the presence of other genes.

- **Recessive**: Gene whose phenotypic expression is masked when in the presence of a dominant allele.

- **Heterozygous**: An animal that carries a gene with two different alleles (Rr).

- **Homozygous**: An animal that carries a gene with two identical alleles (RR or rr).

Punnett Square

- A diagram that is used to predict an outcome of a particular cross or breeding experiment.

- Named after Reginald C. Punnett.

- Used by biologists to determine the probability of an offspring's having a particular genotype and phenotype.

- A tabular summary of every possible combination of one maternal allele with one paternal allele for each gene being studied in the cross.

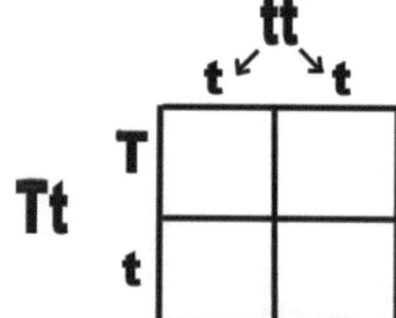

Pea Plants Characteristic to Study: Stem Length

- Using some of Gregor Mendel's work.

    o Assume the dominant allele is tall (T).

    o Assume the recessive allele is dwarf (t).

- Phenotype: The plant is either tall or it's a dwarf.

- Genotypes:

    o Homozygous genotypes:

        ✓ TT with a tall phenotype.

Chapter Resource

- ✓ tt with a dwarf phenotype

- Heterozygous genotype:
    - Tt with a tall phenotype; (since T is dominant over the recessive t)

- Homozygous cross:
    - Homozygous tall (TT)
    - Homozygous dwarf (tt)
        - ✓ Each parent will contribute one allele to the offspring. Since each parent is homozygous, only one form of an allele in a gamete can come from each parent plant.
    - Set up Punnett square:

        Parent TT crossed with Parent Tt

Gametes	T	T
t	Tt	Tt
t	Tt	Tt

    - All offspring have a genotype of Tt and a tall phenotype.

- Heterozygous Cross:
    - Parent (Tt) and parent (Tt)
    - Each parent has two different alleles. Due to Mendel's law of independent assortment, half of the time the parent will donate T to the offspring and t the other half of time (1\2 T and 1\2 t) in the gametes. It is purely random, so predictable ratios can be expected.
    - Set up Punnett square:

        Parent Tt crossed with Parent Tt

Gametes	T	t
T	TT	Tt
t	Tt	tt

- Genotypes of offspring:
    - 1/4 TT: 1/2Tt: : 1/4 tt

- or 25% TT; 50% Tt; and 25% tt
- Genotypic ratio: 1:2:1
- Phenotypes of offspring:
  - 1/4 TT tall; 1/2 Tt tall; and 1/4 tt dwarf
  - Or 75% tall and 25% short.
  - Phenotypic ratio: 3:1 ratio

Summary

Gregor Mendel's experiments with pea plants began the understanding of genetic inheritance and the concepts of phenotype and genotype. Phenotype is the physical appearance of the animal or plant because of its genetic makeup. Genotype is the genetic constitution (makeup) of an animal or plant. Punnett squares provide a tabular summary of every possible combination of one maternal allele with one paternal allele for each gene being studied in the cross.

Additional Resources

A Beginner's Guide to Punnett Square
http://youtu.be/Y1PCwxUDTl8

Deciphering the Genetic Code
http://history.nih.gov/exhibits/nirenberg/index.htm

Genotypes and Phenotypes by Bozeman Science
https://www.youtube.com/watch?v=OaovnS7BAoc

Gregor Mendel and the Principles of Inheritance
http://tinyurl.com/22rc6oa

Punnett Square Calculator
http://scienceprimer.com/punnett-square-calculator

Assessment

Take assessment online here: http://tinyurl.com/AnSci-UnderstandingGenotype
Download and print the assessment by scanning this QR code or by going to this URL: http://www.tagmydoc.com/Ch24AnSci

25 Selection

Major Concept
Variability and natural selection work together to change an animal species.

Objectives
- Define evolution
- Define (genetic) variability
- Define natural selection
- Provide examples of how natural selection and genetic variability may be manipulated by humans to produce disease resistant livestock

Key Terms
- Evolution
- Mutation
- Natural Selection
- Variation

Chapter Resource

 Complementary *full color* illustrations, photos, charts and graphs are available by scanning this QR code or by following this URL: http://www.tagmydoc.com/AS25 This digital resource will enhance your understanding of the chapter concepts.

What is Evolution?
- A change in the genetic makeup of a population with time.
  - Example: A population of black spotted cattle – BB – eventually turn into a population of white spotted cattle – bb – over a period of years.

- A Two-Step Process
  - Development of genetic variability in the population.
  - Natural selection of those variations which are favorable for survival.

Genetic Variability

- If a population starts out with the same genetic makeup (not realistic, since each organism in a population is genetically unique) all new genetic material in that population will arise by mutation.

    o A **mutation** is a random inheritable event.

    o For example, the black spotted cattle; for some reason, the chromosomes that code for black spots are mutated.

- Variation in the chromosomes also occurs when an egg and a sperm unite – both carry unique traits from the parents into one individual – forming an individual which varies from either parent.

    o This **variation** is normally a random event (especially in humans)

    o In the example, the two mutated animals mate with one another (by chance).

    o Using the Punnett Square, here is their offspring:

Chapter Resource

	B	b
B	BB (black)	Bb (black)
b	Bb (black)	bb (white)

Natural Selection

- In a population, certain individuals may be genetically better suited for the environment than others.

- If the environment is such that it favors these individuals, then these are the individuals that will survive to have offspring and pass their genetic makeup to their offspring.

- If the environment does not again change, then these individuals will become the normal population.

- The sequence of events that lead to a certain characteristic being selected by the environment is called natural selection.

    o In this example, a scenario is designed to demonstrate natural selection:

        ✓ Fifty years have gone by since the first mating of mutated cattle, and over the years the once all black spotted herd is now about 75% black-spotted cattle and 25% white spotted cattle.

        ✓ A new disease has also mutated during this time – a disease known as *Bovine Black Spotitis*. This disease is not fatal, but it renders any black spotted bull or cow that contracts the disease sterile.

        Chapter Resource

        ✓ The disease spreads quickly through the herd and within 6 months' time all black spotted cattle are sterile – what will eventually happen to the herd? (Barring death of the white spotted cattle, they will be the only cattle to reproduce, eventually producing a completely white spotted herd).

Use of Genetic Variability and Natural Selection in Producing Improved Livestock

- Livestock managers can take advantage of natural genetic variability in livestock.

    o Brahman cattle (*Bos indicus*) are known for their heat tolerance, because they have adapted to the hotter climate of Asia and Africa.

        ✓ *Bos taurus* does not have this characteristic.

    o As a livestock manager, the role of **natural selector** is taken on by intentionally mating Brahman cattle to European cattle (like the Angus) by determining the mating that will take place (could be called unnatural selection).

        ✓ This mating produces the Brangus breed.
        ✓ Has better carcass characteristics than the Brahman.
        ✓ Has better heat resistance (and calving ease) than the Angus.
        ✓ A similar mating between Brahman and Shorthorn cattle produced the Santa Gertrudis breed.

Summary

A mutation is a random inheritable event. Variation in the chromosomes also occurs when an egg and a sperm unite - both carry unique traits from the parents into one individual - forming an individual which varies from either parent. This variation is normally a random event. The sequence of events that leads to a certain characteristic being selected by the environment is called natural selection.

Additional Resources

Taylor, R.E. and T.G. Field. 2010. Scientific farm animal production: Introduction to animal science. 10th ed. Upper Saddle River, NJ: Prentice Hall.

Khan Academy – Natural Selection
http://tinyurl.com/mtymraj

Natural Selection at Work
http://evolution.berkeley.edu/evolibrary/article/0_0_0/evo_26

Assessment

 Take assessment online here: http://tinyurl.com/AnSci-Selection
Download and print the assessment by scanning this QR code or by going to this URL: http://www.tagmydoc.com/Ch25AnSci

26 Mutation

Major Concept
Genetic mutations occur in populations of animals and these can have beneficial or detrimental effects on animals.

Objectives
- Define mutation
- List three possible causes of a genetic mutation

Key Terms
- Evolution
- Hereditary
- Mutation

Chapter Resource

 Complementary *full color* illustrations, photos, charts and graphs are available by scanning this QR code or by following this URL: http://www.tagmydoc.com/AS26 This digital resource will enhance your understanding of the chapter concepts.

Mutation Defined

- A sudden change in the characteristics of an organism due to a change in the chemical structure of the DNA.

    - This change must be capable of being transmitted faithfully (without change) to future generations; it is inheritable.

- Causes of a mutation include:

    - An error in DNA replication.

    - Exposure to radiation.

    - Exposure to certain toxic (poisonous) chemicals.

- o Other genetic abnormalities such as extra chromosomes or chromosomes crossing or breaking.

Chapter Resource

- Mutation and Evolution

  - o Mutations are recognized as the primary source of the **hereditary** (passing to offspring through genes) variations that make **evolution** (gradual development into a different form) possible.

    - ✓ Mutations may be either harmful or useful to a species.

  - o Those mutations that help an animal survive, are continued because it is the survivors that reproduce and pass the traits on to their offspring.

  - o Those mutations that are harmful to a species are less likely to be passed on to future generations because the animals do not survive to reproduce.

  - o Genetic breeders help to increase the usefulness of mutations by selecting and breeding for traits that are useful to the agricultural producer.

Examples of Mutations Useful to Commercial Agriculture

- A naturally occurring mutation or change occurred in the horse.

  - o Over the years, the horse's hoof characteristic changed to suit a changing environment. The one hoofed horse today is much more useful to agriculture than its smaller four toed counterpart, eohippus, which lived long ago.

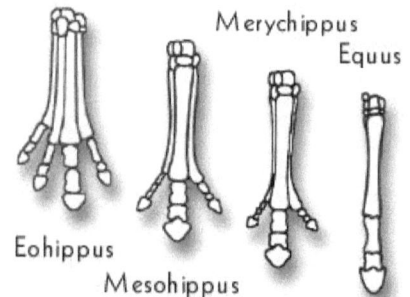

- Another, more current example in livestock, is double muscling which refers to breeds of cattle that carry a mutation repressing the myostatin protein, consequently augmenting muscle growth. Affected breeds include: Belgian Blue, Piedmontese and Parthenais.

- In other instances, man may enhance the rate of change by selecting for certain traits. Two examples are:

  - o Rust Resistant Wheat: Plant breeders have taken advantage of a natural mutation for a strain of rust resistance in wheat. They have bred this trait into other wheat strains to help aid in agriculture production.

  - o High Protein Corn: Plant agronomists and geneticists have continued over the years to always select for the high protein varieties and have tried to breed them exclusively. This selection for the variety (the mutation) producing a higher protein is a "man-made" selection rather than one based solely on "natural selection" (as contrasted to the case with the horse) for desirable traits.

Human-Induced Mutation

- Through controlled use of radiation, chemicals, or physical disruption of the faithful translation of the genetic code (DNA) mutations can be brought about in the laboratory.

    o Goal may simply be to observe and study a process or to try, by chance or design, to improve an animal or plant to make it more beneficial to humans.

- Currently, the use of genetic engineering puts a new twist on human-induced mutations, since technology allows the introduction of specific genes from other animals or even plants into the genetic material of livestock.

Summary

Mutation is a sudden change in the characteristics of an organism due to a change in the chemical structure of the DNA. Causes of a mutation include: An error in DNA replication, exposure to radiation, exposure to certain toxic (poisonous) chemicals and other genetic abnormalities such as extra chromosomes or chromosomes crossing or breaking. Mutations are recognized as the primary source of the hereditary variations that make evolution possible. Plant agronomists and geneticists have continued over the years to always select for the high protein varieties and have tried to breed them exclusively. Mutations may be either harmful or useful to a species. Through controlled use of radiation, chemicals, or physical disruption of the faithful translation of the genetic code (DNA) mutations can be brought about in the laboratory.

Additional Resources

Chernobyl Mutations in Humans and Animals
https://chernobylguide.com/chernobyl_mutations/

Genetics Home Reference
http://ghr.nlm.nih.gov/handbook/mutationsanddisorders/genemutation

Khan Academy: Introduction to Evolution and Natural Selection
http://tinyurl.com/ah4oph2

Assessment

 Take assessment online here: http://tinyurl.com/AnSci-Mutation
Download and print the assessment by scanning this QR code or by going to this URL: http://www.tagmydoc.com/Ch26AnSci

27 Digestion and Absorption

Major Concept

Develop an understanding of the digestive systems in ruminants and non-ruminants and how each part contributes to digestion.

Objectives

- List contributions of microbial digestion (in ruminants) to the host, including synthesis of amino acids and B-vitamins
- Develop an understanding of how feeds are digested and absorbed in the animal's body
- Develop an appreciation for the importance of feeding consistency in animal management

Key Terms

- Abomasum
- Cecum
- Hardware stomach
- Monogastric
- Non-ruminant
- Omasum
- Papillae
- Ptyalin
- Reticulum
- Rumen
- Ruminant

Chapter Resource

Complementary *full color* illustrations, photos, charts and graphs are available by scanning this QR code or by following this URL: http://www.tagmydoc.com/AS27 This digital resource will enhance your understanding of the chapter concepts.

Digestive Systems

- In its simplest form, the digestive system is a tube extending from the mouth to the anus with associated organs.

    o This includes mouth, esophagus, stomach, intestines, anus and other associated organs like the liver, teeth, pancreas and salivary glands.

    o Digestive systems vary according to whether the animals are herbivores, carnivores or omnivores.

- Types of digestive systems are ruminant or non-ruminant or single stomach.

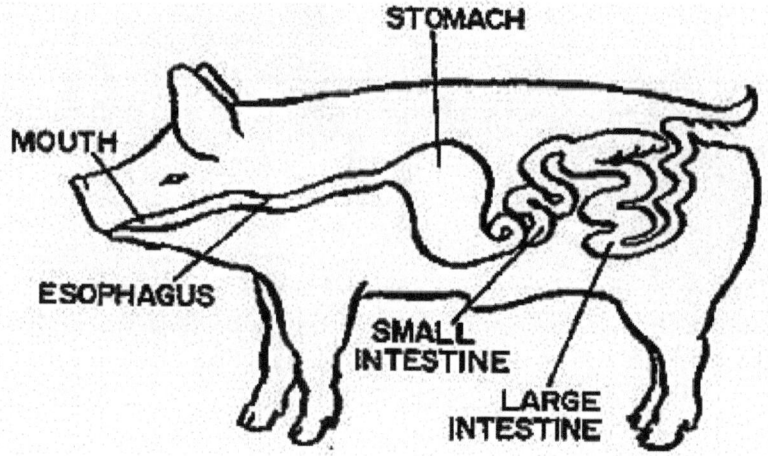

Ruminant Defined

- An animal with four distinct compartments in its stomach, which swallows its food essentially unchewed, regurgitates it, chews it thoroughly and reswallows it.

    o Examples include cattle, sheep, goats

Chapter Resource

Non-ruminant (Monogastric) Defined

- An animal, having a single compartment in its stomach, which swallows its food after chewing and does not regurgitate.

    o Examples include pigs, humans, bears and dogs.

Four Compartments of the Ruminant

1. **Reticulum** – honeycomb – 5% of capacity
2. **Rumen** – paunch – 80% of capacity
3. **Omasum** – manyplies – 7% of capacity
4. **Abomasum** – true stomach – 8% of capacity

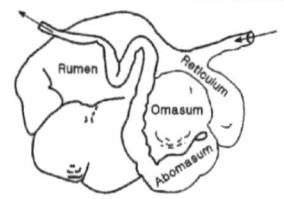

Functions of the Compartments

- Listed in the order that feed passes through them:

- Reticulum (nicknamed the **hardware stomach**):

    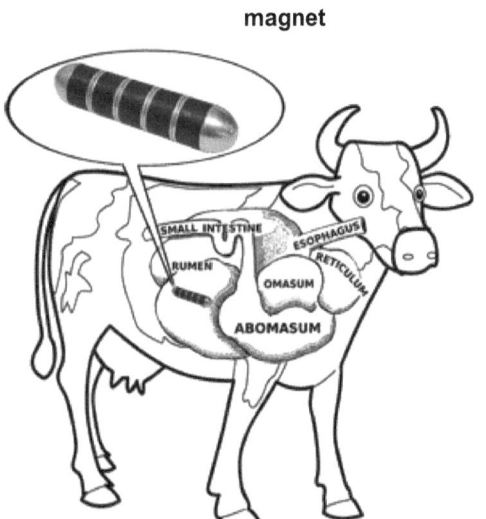

    - To furnish additional storage space and hold foreign materials such as nails and wire which may cause serious damage to the other body organs if they pass through the digestive system.

    - Often a magnet is fed to ruminants to catch and hold this foreign material in the reticulum.

- Rumen

    - Largest compartment; capacity of 40 to 60 gallons in mature cattle.

    - Serves as a storage area for feed.

    - Millions of bacteria and protozoa break down the feed in the rumen.

    - Bacterial action is the reason ruminants can digest large quantities of roughage and live on a much lower protein diet than a monogastric.

    - The rumen does not develop in the young until they begin to eat solid food (as they are weaned).

- Omasum

    - A muscular section that squeezes out the water from the feed before it enters the abomasum

    Chapter Resource

- Abomasum (the true stomach)

    - Digestive juices begin chemical changes to break down the proteins, carbohydrates and fats into simpler substances which can pass into the small intestine for further digestion and absorption into the blood stream.

- Regurgitation

    - During the process of eating, ruminants chew their feed just enough to make swallowing possible. After consuming their feed, it is brought up from the rumen and chewing is completed.

Non-ruminant Digestion

- Food is swallowed directly into the single compartment stomach, where it is mixed with the digestive juices. Because there is very little bacterial action, there is no conversion of low-quality protein to high quality protein.

- **Monogastrics** are unable to digest large quantities of fiber unless they have an enlarged cecum (examples include horses, rabbits and guinea pigs).

    - The **cecum** is a blind pouch or cul-de-sac at the first portion of the large intestine.

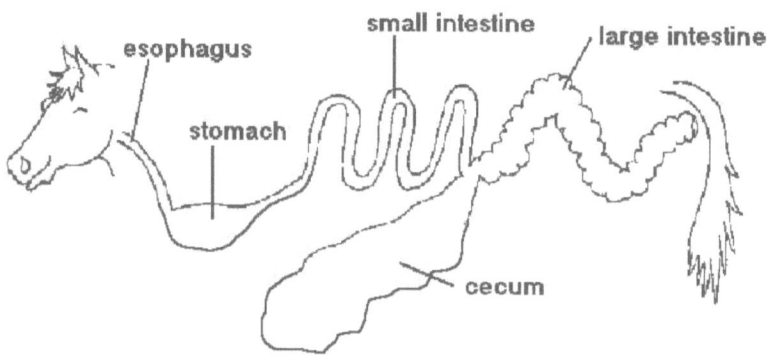

    - While the rumen is the main place for bacterial breakdown in ruminants, it happens in the large intestine and particularly in the cecum of the horse and rabbit.

Understanding Digestion and Absorption

- Digestion

    - Process by which large, complex food molecules are broken down into simpler molecules within the digestive tract to prepare them for absorption and use in the body.

Chapter Resource

- Enzymes

    - Organic compounds which bring about changes in other organic compounds without themselves being changed or broken down.

    - Help in breaking down complex feeds into simpler forms.

- Importance of chewing (example: kernel of corn)

- o If a kernel of corn escaped chewing when first eaten and was not brought back up during rumination, the kernel may pass through the entire digestive tract in an unbroken condition.

- o Increase the surface area so enzymes can act on the food and break it down.

- Digestion of carbohydrates

  - o Carbohydrates must first be broken down into simple starches and sugars before they can be absorbed in the small intestine.

  - o Carbohydrate digestion begins in the mouth with those animals whose saliva contains the enzyme **ptyalin** (i.e., non-ruminants).

  - o In ruminants, whose saliva does not contain ptyalin, carbohydrate digestion begins in the rumen.

  - o The rest of the carbohydrate is digested and absorbed in the small intestine.

- Bacterial digestion of carbohydrates (including cellulose) and protein

  - o Millions of bacteria and protozoa grow in the mass of water and feed in the rumen.

  - o These organisms secrete enzymes which digest the carbohydrates and proteins.

- Protein digestion

  - o Proteins are broken down into smaller units, including amino acids (building blocks of protein) in the stomach (abomasum in ruminants) and in the small intestine.

- Fat digestion

  - o Fats do not undergo digestion until they reach the small intestine.

  - o Bile salts and enzymes from the pancreas help in the breakdown of fats so they can be absorbed in the small intestine.

- Villi

  - o Line the intestines and are minute finger-like processes or **papillae**.

  - o Provide a larger surface area in the lining of the intestine so there is more surface area for absorption of nutrients

Summary

A ruminant is an animal with four distinct compartments in its stomach. A non-ruminant (monogastric) is an animal, having a single compartment in its stomach. The four compartments of the ruminant are reticulum, rumen, omasum and abomasum. In non-ruminant digestion, the food is swallowed directly into the single compartment stomach, where it is mixed with the digestive juices. Monogastrics are unable to digest large quantities of fiber unless they have an enlarged cecum.

Additional Resources

Taylor, R.E. and T.G. Field. 2011. Scientific farm animal production. 10th ed. Upper Saddle River, NJ: Prentice Hall.

Digestive Physiology
http://www.ag.auburn.edu/~chibale/an02physiology.pdf

The Cow's Digestive System
http://animalscience.tamu.edu/wp-content/uploads/sites/14/2012/04/nutrition-cows-digestive-system.pdf

Assessment

Take assessment online here: http://tinyurl.com/AnSci-Digestion
Download and print the assessment by scanning this QR code or by going to this URL: http://www.tagmydoc.com/Ch27AnSci

28 Nutrient Needs

Major Concept

Livestock require basic nutrients in their diet for successful and efficient production.

Objectives

- List the six classes of nutrients and identify the importance of each
- Give examples of the feeds contained in each nutrient

Key Terms

- Digestible nutrient
- Nutrient
- Nutrition
- Osmosis
- Steroids

Chapter Resource

 Complementary *full color* illustrations, photos, charts and graphs are available by scanning this QR code or by following this URL: http://www.tagmydoc.com/AS28 This digital resource will enhance your understanding of the chapter concepts.

Nutrient Needs Definitions

- Object of nutrition

  o To obtain and use feedstuffs and convert them to desirable products such as meat, milk, eggs, fiber and work.

- Definition of **nutrition**

  o Science dealing with the utilization of feed/food by the body and all body processes which transform feed/food into body tissues and activities.

- Definition of **nutrient**

  o A substance that provides nourishment essential to physiological processes involved in growth and the maintenance of life.

- Definition of **digestible nutrient**

- That portion of a nutrient which may be broken down (digested) and absorbed and used by the body.

Six General Classes of Nutrients (Needed by all livestock and humans):

1. Protein
2. Carbohydrates
3. Fats
4. Minerals
5. Vitamins
6. Water
7. Air (sometimes considered the seventh nutrient)

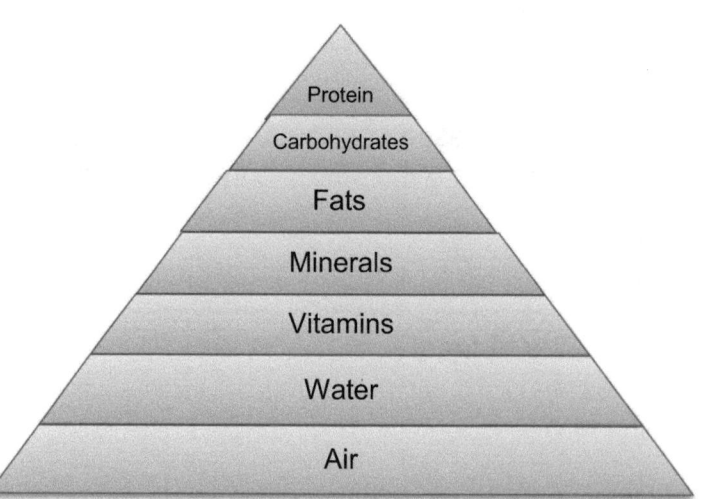

- Water
  - Largest constituent of plants and animals.
  - Water accounts for 70% or more of the composition of most plants and animals.
  - Functions of water in the body are:
    - ✓ Controls body temperature.
    - ✓ Enables living plants and animals to hold their shape.
    - ✓ Is involved in the transport of nutrients at the cell level through **osmosis** (across a biological membrane) and diffusion.
    - ✓ Helps in the digestion of feeds.
    - ✓ Serves as a carrier for waste-products resulting from body functions.
    - ✓ Is a major (by volume) part of all body fluids.

- Carbohydrates
  - Furnish energy for body functions, growth, fattening, reproduction, etc.
  - Represent the largest part of an animal's feed supply.
  - Usually the fibrous part of the diet.

- Include sugars, starch and cellulose.
- Composed of carbon (C), oxygen (O) and hydrogen (H).

- Fats
    - Furnish a concentrated source of energy, up to 2.25 times as much energy per unit of weight, as do carbohydrates.
    - Form cholesterol, **steroids** (including some hormones) and other body compounds.
    - Found in every cell in the body.
    - When absent from the diet, affects (among others) the condition of the skin and hair.
    - Composed of carbon (C), oxygen (O) and hydrogen (H), but contain much larger proportions of carbon and hydrogen than do carbohydrates.
    - Other functions:
        - ✓ Energy reserves
        - ✓ Protection for vital organs
        - ✓ Insulate the body

Chapter Resource

- Proteins
    - Provide essential amino acids.
    - Are essential in livestock feeding because they are needed throughout life for growth and repair.
    - Helps to form the greater part of muscles, internal organs, skin, hair, wool, feathers, hoofs and horns.
    - Contains carbon (C), hydrogen (H) and oxygen (O) (in common with fats and carbohydrates) but also contains a fairly constant percentage of nitrogen (N) (about 16 %).

- Minerals
    - Are needed in nearly all parts of the body but are found primarily in bones and teeth.

- Make up important parts of many organic materials including blood (iron (Fe) in the hemoglobin molecule in the red blood cell which carries oxygen).
- Affect heartbeat, which depends upon mineral balance to maintain its regularity (minerals involved are calcium (Ca), sodium (Na) and potassium (K)).
- Involved in nerve transmission.
- Divided into two groups based on amounts needed by the body:
    1. Macrominerals
    2. Microminerals (or trace minerals)

- Vitamins

Chapter Resource

- Needed only in minute amounts, but are essential for life and health
- Divided into two groups:
    1. Fat-soluble
    2. Water-soluble

Summary

Object of nutrition is to obtain and use feedstuffs and convert them to desirable products such as meat, milk, eggs, fiber and work. The definition of nutrition is the science of dealing with the utilization of food/feed by the body and all body processes which transform food/feed into body tissues and activities. A nutrient is any substance that provides nourishment essential to physiological processes involved in growth and the maintenance of life. Six major classes of nutrients are required by livestock: protein, carbohydrates, fats, minerals, vitamins and water. Water represents the largest constituent of plants and animals.

Additional Resources

Taylor, R.E. and T.G. Field. 2011. Scientific farm animal production. 10th ed. Upper Saddle River, NJ: Prentice Hall.

Basic Nutrient Requirements of Beef Cows
http://edis.ifas.ufl.edu/an190

Livestock Nutrition, Husbandry, and Behavior
http://directives.sc.egov.usda.gov/OpenNonWebContent.aspx?content=17741.wba

Six Classes of Nutrients: Unit 6: Productivity and Nutrition
http://ag.ansc.purdue.edu/nielsen/www245/lecnotes/Nutrition.html

Assessment

 Take assessment online here: http://tinyurl.com/AnSci-NutrientNeeds
Download and print the assessment by scanning this QR code or by going to this URL: http://www.tagmydoc.com/Ch28AnSci

29 Protein Needs

Major Concept

Protein in the diet provides amino acids to the animal to be used in growth and other processes.

Objectives

- List two factors to consider when feeding protein
- Identify the essential amino acids
- Name two factors affecting the requirement for protein in an animal
- Identify the animals that can use urea as a protein substitute

Key Terms

- Amino acids
- Germ
- Urea

Chapter Resource

 Complementary *full color* illustrations, photos, charts and graphs are available by scanning this QR code or by following this URL: http://www.tagmydoc.com/AS29 This digital resource will enhance your understanding of the chapter concepts.

Protein Needs

- Two most important factors to remember when feeding protein:

    1. Amount of protein
    2. Quality or kind of protein

Excess Protein

- There is no danger in feeding a larger amount than animals require; however, protein is usually the most expensive part of the feed, so overfeeding is impractical.

- o Once an animal has consumed the needed amount of protein for cell construction, muscle, fetal growth, etc., the rest is broken all the way down for use as body energy.

- o Carbohydrates usually a much less expensive source of energy.

Essential Amino Acids

- Those which cannot be made in the body from other substances, or which cannot be made in sufficient amounts for physiological (body function) needs.

- There are nine or ten essential **amino acids** (depending on the source of information).

Essential amino acids	Non-essential amino acids
Histidine	Alanine
Isoleucine	Arginine
Leucine	Asparagine
Lysine	Aspartic acid
Methionine	Cysteine
Phenylalanine	Glutamic acid
Threonine	Glutamine
Tryptophan	Glycine
Valine	Proline
	Serine
	Tyrosine

Amino Acid Requirements of Animals

- Kind of animal

- Body functions of the animal including age, work, lactation and fetal growth.

Chapter Resource

- o Certain amino acids necessary for growth are not essential for merely maintaining an animal.

Protein in the Rations of Ruminants and Non-Ruminants:

- All nine or ten essential amino acids can be made in ruminants by the rumen bacteria from simple forms of nitrogen in the feed. (Therefore, the bacterial protein may provide all the essential amino acids, even though they are lacking in the feed which the ruminant eats).

- The **germ** is the part of the grain kernel which usually contains available protein.

- **Urea** ($CO(NH_2)_2$) (Note: Say 1 part carbon, 1 part oxygen, 2 parts nitrogen and 4 parts hydrogen) is often used as a protein substitute in ruminants. It is a source of nitrogen which the rumen "bugs" can use to make bacterial protein.

    o Urea is used ONLY in ruminant rations.

- Protein of animal origin:

    o Examples include fish meal, meat and bone meal, dried milk products, blood meal and tankage (from "rendered down" animals).

- Protein of vegetable (legume or seed) origin:

    o Examples include soybean oil meal, soybeans, cottonseed meal, linseed meal, legumes.

- Young vs. mature animals:

    o Young animals are building new tissue as they grow and have higher protein needs.

    o Other times in an animal's life when there is a need for increased proteins include lactation and pregnancy.

Chapter Resource

Summary

The two most important factors to remember when feeding protein are the amount of protein and the quality or kind of protein. There are ten essential amino acids. Amino acid requirements of animals depend on the kind of animal and body functions of the animal including age, work, lactation and fetal growth. The germ is the part of the grain kernel which usually contains available protein. Other times in an animal's life when there is a need for increased proteins, include lactation and pregnancy.

Additional Resources

Morrison, F.B. 1962. Feeds and feeding. 9th ed. Morrison Pub. (pg. 48-57).

Taylor, R.E. and T.G. Field. 2011. Scientific farm animal production. 10th ed. Upper Saddle River, NJ: Prentice Hall.

Grain and Protein Supplements for Beef Cattle on Pasture
http://extension.missouri.edu/p/G2072

Nutrition and Management: Types and Sources of Protein
http://www1.agric.gov.ab.ca/$department/deptdocs.nsf/all/beef11678

Assessment

 Take assessment online here: http://tinyurl.com/AnSci-ProteinNeeds
Download and print the assessment by scanning this QR code or by going to this URL: http://www.tagmydoc.com/Ch29AnSci

30 Carbohydrates and Fats

Major Concept

Carbohydrates and fats are important components of livestock feed.

Objectives

- Name the three general classes of carbohydrates and provide examples of each
- Identify the elements that make up carbohydrates and fats
- List the functions of carbohydrates and fats in the diet

Key Terms

- Carbohydrates
- Cellulose
- Glycogen
- NFE (Nitrogen-Free Extract)
- Rancid

Chapter Resource

Complementary *full color* illustrations, photos, charts and graphs are available by scanning this QR code or by following this URL: http://www.tagmydoc.com/AS30 This digital resource will enhance your understanding of the chapter concepts.

Carbohydrates Needed in Livestock Feed

- **Carbohydrates** are:

    o Compounds composed of carbon (C), hydrogen (H) and oxygen (O). Examples include starches and sugars.

    o Form about 75% of all dry matter in plants.

    o "Chief source of energy" in animal feed.

- Three classes of carbohydrates are:

    1. Sugars
    2. Cellulose (crude fiber)
    3. Starches, or glycogen if stored in the body

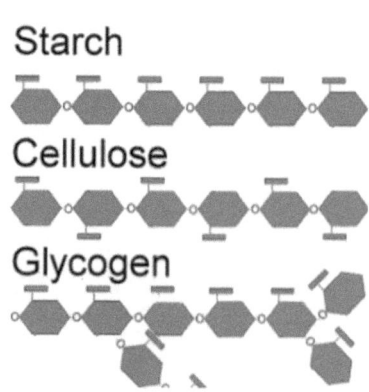

- Feeding value of **cellulose**:
  - Hard to digest - digested only through the action of bacteria in the rumen or cecum and even then, not as completely as other carbohydrates.

- Crude fiber in the diet:
  - Includes relatively indigestible material such as cellulose and other complex carbohydrates.
  - Found mostly in hay and other "rough" or "woody" plants.

- **NFE** (Nitrogen-Free Extract) of a feed analysis:
  - Includes starch, the sugars and the more digestible portions of the carbohydrates in a feed.

- Sources of carbohydrates:
  - Grains – corn, oats, barley, wheat, grain sorghums
  - Roughages – legume hay, corn silage
  - Molasses

Chapter Resource

Fats in Livestock Feeding

- Compounds composed of carbon (C), hydrogen (H) and oxygen (O).

- Fats vs. Carbohydrates:
  - Fats furnish 2.25 times as much energy per pound (per unit of weight) as carbohydrates.

- Difference between fats and oils:
  - Fats are solid at ordinary (room) temperatures while oils are liquid.

- Fats are the ether extract (EE) portion of feed composition.

- Function of fat in the diet:
  - Provides fat- soluble vitamins like A, D, E and K.
  - Aids in absorption of calcium.

- o Increases **palatability** (acceptable to taste) of a ration.
- o Decreases dustiness and dryness of ration.
- o Increases production energy of ration.
- o Needed as a component in every cell in the body.

Chapter Resource

- **Rancid** (spoiled) fats create problems such as:
  - o Decreasing palatability of a ration, causing it to have a stale and unpleasant odor.
  - o Destroying certain fat-soluble vitamins.
  - o Producing soft lard in hogs.
  - o Decreasing milk production.

- Dietary requirement of fat in farm animals is low; for example, the maximum amount is about 3% in dairy rations.

- High fat content in rations:
  - o When rations are high in fat – protein and water soluble vitamin content needs to be increased because animals limit their feed intake by satisfying their caloric requirement sooner. (Animals don't eat enough to obtain necessary proteins, vitamins, etc. unless the feed is "fortified.")
  - o Danger that young growing animals on high fat rations may not get sufficient protein for maximum growth unless the protein is increased to higher level or greater attention is given to protein quality and/or the amount of fat is reduced.

- Sources of fat:
  - o Lard
  - o Tallow
  - o Vegetable oils
  - o Soybean and peanut oil
  - o Feed grains

Summary

Carbohydrates and fats are composed of carbon (C), hydrogen (H) and oxygen (O). Carbohydrates form about 75% of all dry matter in plants and are the "chief source of energy" in animal feed. The three classes of carbohydrates are sugars, starches and crude fiber. Crude fiber includes relatively indigestible material such as cellulose and other complicated carbohydrates. NFE (Nitrogen-Free Extract) includes starch, the sugars, and the more digestible portions of the carbohydrates in a feed. Fats furnish 2.25 times as much energy per pound as carbohydrates. Fats aid in the absorption from the food of vitamin A and especially of carotene, help in the absorption of calcium, increases the palatability of a ration, decreases the dustiness and dryness of the ration, increases the production energy of the ration and is found as a component in every cell in the body.

Additional Resources

Parker, R. 2008. Equine science. 3rd ed. Clifton Park, NY: Thomson Delmar Learning. (Pg. 231-232).

Taylor, R.E. and T.G. Field. 2011. Scientific farm animal production. 10th ed. Upper Saddle River, NJ: Prentice Hall.

Energy in Beef Cattle Diets
http://extension.msstate.edu/node/7016

Productivity and Nutrition
http://ag.ansc.purdue.edu/nielsen/www245/lecnotes/Nutrition.html

Assessment

Take assessment online here: http://tinyurl.com/AnSci-Carbohydrates
Download and print the assessment by scanning this QR code or by going to this URL: http://www.tagmydoc.com/Ch30AnSci

31 Vitamins and Minerals

Major Concept

Vitamins and minerals play essential roles in livestock feeding.

Objectives

- Name the two categories of vitamins
- Identify the vitamins and their functions
- Name the two categories of minerals and at least three minerals in each category
- Identify the minerals needed by livestock and their function

Key Terms

- Dermatitis
- Essential
- Fat-soluble vitamins
- Macrominerals
- Microminerals
- Synthesized
- Scurvy
- Water-soluble vitamins

Chapter Resource

 Complementary *full color* illustrations, photos, charts and graphs are available by scanning this QR code or by following this URL: http://www.tagmydoc.com/Ch31ABCombined This digital resource will enhance your understanding of the chapter concepts.

Vitamins and Their Functions in Livestock Feeding

- Vitamins

  o **Essential** (necessary) organic nutrient which is required in minute amounts.

  o Cannot be made (**synthesized** in the body) and must be obtained through an external source such as diet, action of sunshine on skin (vitamin D) or synthesized in the rumen, cecum or large intestine by bacteria.

  o Required for growth, maintenance, reproduction and lactation.

- Vitamins vs. other nutrients

- Not used to directly build body tissue, instead they are components of specific enzyme and hormone systems which affect all bodily activities and are thus essential for normal life process.

- Two Classes of Vitamins – Fat-soluble and water-soluble:

  - **Fat-soluble vitamins** (any vitamin that is soluble in fats) can be stored and accumulated in the liver and other fatty tissues.

    ✓ Generally, provided in the fat component of a feedstuff

    ✓ Only very limited amounts of water-soluble vitamins are stored.

  - **Water-soluble vitamins** (carried to the body's tissues but are not stored in the body) must be provided regularly in the ration in adequate amounts. Water-soluble vitamin deficiencies can develop in a short period of time.

    ✓ Fat-soluble vitamins include A, E, D and K

    ✓ Water-soluble vitamins include C, B vitamins and choline

Chapter Resource

- Vitamin Deficiencies

  - Most animals on commercially prepared feeds will not develop vitamin deficiencies because the feeds are usually fortified and balanced.

  - Animals on pasture or range are the ones which typically develop deficiencies.

    ✓ This may be due to a seasonal deficiency in the feed, a toxin (poison) which blocks absorption of a vitamin.

Function, Deficiency Signs and Sources

- Vitamin A

  - Function – Development of normal, healthy, epithelial tissue or skin and nerve tissue which aids in building up resistance to infection. All animals require a source of Vitamin A; especially important in the ration of pregnant females.

  - Deficiency Signs – Retarded growth in the young and the development of a peculiar condition around the eyes known as Xerophthalmia. Night blindness and affects reproduction.

  - Sources – Carotene (vitamin A precursor), animal body oils (cod fish and tuna), legume forages and can be synthetically produced.

- Vitamin E
  - Function – Normal reproduction, serves as the protector of vitamin A in the diet of poultry and cattle.
  - Deficiency Signs – "Crazy chick disease," white muscle disease in ruminants and stiff lamb disease (affects the nerves and muscles).
  - Sources – Synthetic tocopherol for poultry, cereal grains, wheat germ oil, forages, protein concentrates and alpha-tocopherol.
    - ✓ Feeding of rancid or spoiled fats may cause white muscle disease.

- Vitamin D
  - Function – Essential for the proper use of calcium and phosphorus to produce normal, healthy bones.
  - Deficiency Signs – Retarded growth, misshapen bones (rickets) and lameness.
  - Sources – Forage crops, fish liver oils, irradiated yeast. Chemical forms: Vitamin D2 and Vitamin D3.

- Vitamin K

  Chapter Resource

  - Function – Necessary for the maintenance of normal blood coagulation.
  - Deficiency Signs – Blood loses its power to clot and serious hemorrhages can result from slight wounds or bruises.
  - Sources – Leafy forages, fish meal, liver, soybeans and the synthetic compound – menadione.

- Vitamin C (Ascorbic acid)
  - Function – Influences the metabolism of calcium in the body. (Not required in rations of farm animals).
  - Deficiency Signs – **Scurvy** (swollen and painful joints and bleeding gums in humans) and brittleness of bones.
  - Sources – Citrus fruits, tomatoes, leafy vegetables and potatoes.

- Vitamin B_1 (Thiamin)
  - Function – Required for the normal metabolism of carbohydrates.

- o Deficiency Signs – Loss of appetite, muscular weakness, severe nervous disorders, general weakness and wasting (BeriBeri in humans).

- o Sources – Raw, whole grains and especially their seed coats and embryos; fresh green forage; yeast.

- Vitamin B_2 (Riboflavin)

  - o Function – Necessary for normal embryo development, important in the metabolism of amino acids and carbohydrates.

  - o Deficiency Signs – Poor reproduction characterized by small litters and deformed young (cleft palate and club-footedness). Curly toe paralysis in chicks, digestive disturbances, general weakness, eye trouble.

  - o Sources – Milk and dairy by-products, yeast, green forages, well cured hay, whole grains, wheat bran and synthetic riboflavin.

Chapter Resource

- Niacin

  - o Function – Recognized and designed as the pellagra preventing vitamin or black tongue factor. (Ruminants do not need niacin because they produce it through bacterial syntheses in the rumen).

  - o Deficiency Signs – Reddening of the skin and development of sores in the mouth and intestinal tract resulting in bloody diarrhea. Fowl - slipped tendons or perosis, poor feathering.

  - o Sources – Dried yeast, rice bran, wheat bran, peanut oil meal, green forage, barley grain, sorghum grains, fish meal, meat scrap and the amino acid tryptophan.

- Vitamin B_6 (Pyridoxine)

  - o Function – Deals with fat metabolism and seems to be associated with the transportation and synthesis of unsaturated fatty acids.

  - o Deficiency Signs – Specific **dermatitis** (inflammation and redness of skin) and convulsions.

  - o Sources – Cereal grains, milk, cane molasses, yeast and rice polish. (No danger of a deficiency of this vitamin in animal rations).

- Pantothenic Acid

  - o Function – Plays an essential role in many basic biochemical reactions.

- ✓ Synthesized by bacterial action in the rumen.
  - o Deficiency Signs – Abnormal skin condition about the face and eyes, retarded growth, and poor feather development. For example, in swine – poor appetite, slow growth, coughing, diarrhea, dermatitis and stilted gait.
  - o Sources – Fresh small grains, alfalfa hay, green pasture, wheat bran, peanut oil, dairy by-products and yeast

- Vitamin B_{12}
  - o Function – Essential for normal growth, reproduction and blood formation.
  - o Deficiency Signs – Poor growth and reproduction.
  - o Sources – Chemically pure substance B_{12}, fish meal, liver meal and dried milk products.

- Choline (is included with the vitamins because it is essential and a dietary source is needed; however, it is synthesized in the body to a limited extent and therefore does not fit the definition of a true vitamin).
  - o Function – Helps with the transportation and use of fatty acids.
  - o Deficiency Signs – Kidney and liver damage, slipped tendons or perosis in chicks, and development of fatty livers.
  - o Sources – Choline chloride in ration, liver meal, brewer's yeast, fish meal, cottonseed meal, and soybean oil meal.

Chapter Resource

- Biotin (considered a "B-Complex" vitamin)
  - o Function – In chicks: prevents slipped tendons and increases the hatchability of eggs.
  - o Deficiency Signs – Slipped tendons in chicks, and reduces the hatchability of eggs.
  - o Sources – Ordinary feeds.

- Folic Acid (Folacin) (considered a "B-complex" vitamin)
  - o Function – Required for normal blood cell development and is considered an anti-anemia vitamin.

- o Deficiency Signs – Megaloblastic anemia, retarded growth, poor feathering, bleaching of feather and poor hatchability of eggs.

- o Sources – Forages, oil meals and cereal grains.

Minerals and Their Functions in Livestock Feeding

- Required Minerals:

  - o As many as 20 minerals may be required by the animal body; generally, 14 are considered important in feeds and rations.

- Two minerals needed most Calcium (Ca) and Phosphorus (P).

  - o Skeleton on an animal's body is chiefly of these two minerals.

- Based on the relative amount needed by mammals, minerals are divided into two categories as follows:

  - o **Macrominerals** (more needed): Calcium (Ca), Phosphorus (P), Potassium (K), Sodium (Na), Chlorine (Cl), Magnesium (Mg) and Sulfur (S).

  - o **Microminerals** (less needed): Iodine (I), Copper (Cu), Iron (Fe), Manganese (Mn), Zinc (Zn), Molybdenum (Mo) and Selenium (Se).

Function, Deficiency Signs and Sources of Minerals

- Calcium (Ca)

  - o Function – Major component of bones and teeth and essential in blood coagulation.

  - o Deficiency Signs – Retarded growth, deformed bones (rickets) and soft-shelled eggs.

  - o Sources – Milk, oyster shells and limestone

Chapter Resource

- Phosphorus (P)

  - o Function – Essential for the formation of bones, teeth and body fluids. Required for metabolism, cell respiration and normal reproduction.

  - o Deficiency Signs – Like calcium deficiency, lack of appetite, poor reproduction and unthrifty appearance.

- Sources – Dicalcium phosphate, bone meal and low fluorine phosphates

- Sodium-Chloride-Potassium (Na, Cl, K)

  - Function – Required for the formation and retention of body fluids, such as protoplasm, blood and digestive juices.

  - Deficiency Signs – Poor condition and depressed appetite.

  - Sources – Salt supplements and injectable products.

- Iron (Fe)

  - Function – Essential for the function of every organ and tissue of the body; a component of hemoglobin.

  - Deficiency Signs – Nutritional anemia.

  - Sources – Forages and copper salts.

Chapter Resource

- Copper (Cu)

  - Function – Should be present in animal tissues for iron to be properly used.

  - Deficiency Signs – Poor pigmentation of feathers, stringy wool, sway back lambs, lack of muscle coordination and anemia.

  - Sources – Forages and copper salts.

- Cobalt (Co)

  - Function – Required as a nutrient for the microorganisms in ruminants.

  - Deficiency Signs – Lack of appetite, loss of weight, rough hair coat and death in extreme cases.

  - Sources – Legume forages and salt containing cobalt.

- Magnesium (Mg)

  - Function – Similar to calcium and phosphorus.

  - Deficiency Signs – Animals are irritable, their heart beat is irregular; eventually severe kidney damage.

  - Sources – Mineral supplements, ordinary feeds.

- Manganese (Mn)
  - Function – Affects growth and reproduction.
  - Deficiency Signs – Rabbits develop small crooked legs, chicks display a deformity of leg bones, causes sterility and ovulation is irregular.
  - Sources – Salt supplements and ordinary feed.

- Sulfur (S)
  - Function – Necessary in organic form of the essential amino acids, cystine and methionine.
  - Deficiency Signs – None; very little sulfur is needed in rations.
  - Sources – Legumes; also, occurs in the amino acids of proteins.

- Zinc (Zn)
  - Function – Necessary for good growth and for normal hair development, aids in wound healing and serves as an activator of other enzyme systems.
  - Deficiency Signs – Growth and hair development are retarded.
  - Sources – Trace mineralized salts.

Chapter Resource

- Iodine (I)
  - Function – Essential for the formation of the hormone, thyroxine.
  - Deficiency Signs – Nutritional goiter (enlarged thyroid gland), poor growth, impaired development.
  - Sources – Iodized or mineralized salt.

- Fluorine (Fl)
  - Function – Difficult to assign any function but is found in bones, teeth and skin.
  - Deficiency Signs – Tooth decay (also caused by toxic [too high] levels in rations).
  - Sources – Calcium phosphates and mineral supplements.

- Molybdenum (Mo)
  - Function – Role in enzyme systems.
  - Deficiency Signs – Extreme diarrhea with loss of weight and production.
  - Sources – Forages
- Selenium (Se)
  - Function – Can replace sulfur in the amino acids, cystine and methionine; also, a role with vitamin E.
  - Deficiency Signs – Loss of hair from the mane and tail in horses, loss of hair from the tail of cattle, and loss of hair in swine, hoofs slough off, lameness occurs and appetite decreases.
  - Sources – Legumes
- Sources of mineral supplements
  - High quality ground limestone
  - Steamed bone meal
  - Salt
  - Trace mineral mixture
  - Type of roughage: legumes

Chapter Resource

Summary

Vitamins are an essential organic nutrient required in minute amounts. Fat-soluble vitamins can be stored and accumulated in the liver and other fatty tissues, whereas only very limited number of water-soluble vitamins are stored. Disregarding other physiological problems, most healthy animals on commercially prepared feeds will not develop vitamin deficiencies because the feeds are usually fortified and balanced. At least 14 minerals are required by the animal body. Sources of mineral supplements include high quality ground limestone, steamed bone meal, salt and trace mineral mixture, as well as that supplied by other feedstuffs. Minerals function in a variety of body systems but most notably the calcium (Ca) and phosphorus (P) needed for the skeleton.

Additional Resources

Kellems, R.O. and D.C. Church. 2009. Livestock feeds and feeding. 6th ed. Upper Saddle River, NJ: Prentice Hall.

Suttle, N.F. 2010. Mineral nutrition of livestock. 4th ed. Cambridge, MA: CAB, International.

Beef Cattle Mineral Nutrition
http://www.ag.ndsu.edu/pubs/ansci/beef/as1287.pdf

Minerals and Vitamins for Sheep
http://www.sites.ext.vt.edu/newsletter-archive/livestock/aps-06_10/aps-373.html

Mineral and Vitamin Nutrition for Beef Cattle
http://extension.msstate.edu/node/7260

Assessment

 Take assessment online here: http://tinyurl.com/AnSci-Vitamins-Minerals
Download and print the assessment by scanning this QR code or by going to this URL: http://www.tagmydoc.com/Ch31AnSci

32 Feed Composition

Major Concept

Feeds represent a major cost of livestock production.

Objectives

- List the six general groups identified in the chemical composition of feeds
- Define crude protein, crude fiber, concentrates, total digestible nutrients, net energy and forages or roughages
- Identify the nitrogen-free extract portion of feeds

Key Terms

- Acid Detergent Fiber(ADF)
- As-fed
- Bacterial Crude Protein(BCP)
- Calorie
- Concentrates
- Crude Fiber(CF)
- Crude Protein(CP)
- Degradable Intake Protein(DIP)
- Digestible Protein(DP)
- Dry matter
- Forages
- Metabolizable Protein(MP)
- Net Energy(NE)
- Neutral Detergent Fiber(NDF)
- Nitrogen-Free Extract (NFE)
- Roughages
- Total Digestible Nutrients (TDN)
- Undegradable Intake Protein(UIP)

Chapter Resource

Complementary *full color* illustrations, photos, charts and graphs are available by scanning this QR code or by following this URL: http://www.tagmydoc.com/AS32 This digital resource will enhance your understanding of the chapter concepts.

Analysis

- Feeds are classified according to chemical analysis into six general groups and expressed as percentages:

    1. Water/moisture
    2. Mineral matter or ash
    3. Protein
    4. Fiber
    5. Fat
    6. Nitrogen-free extract

- Water is expressed as the percentage of moisture in a feed.

    o Moisture dilutes the concentration of nutrients in a feed.

    o Feed generally express in terms of **dry matter** which is the moisture-free content of the sample.

    o **As-fed** means a feed's nutritional value is being expressed with its typical moisture/water content.

- Mineral/ash measures the total mineral content in a feed but not individual minerals.

    o Other tests on feeds are used to determine the levels of important minerals like calcium (Ca), phosphorus (P), copper (Cu), zinc (Zn), etc.

- Protein contains the nitrogen (N) in the feed and early measurements of protein content were determined by the amount of N in the feed sample.

    o **Crude Protein** (CP) measures the nitrogen content of a feedstuff, including both true protein and non-protein nitrogen.

        ✓ Protein supplies amino acids to livestock, some of which are essential to monogastric animals.

    o **Degradable Intake Protein** (DIP) is the fraction of the crude protein which is degradable in the rumen and provides nitrogen for rumen microorganisms to synthesize **bacterial crude protein** (BCP) which is protein supplied to the animal by rumen microbes.

    o **Undegradable Intake Protein** (UIP): is commonly called "bypass protein" because it bypasses rumen breakdown and is mainly digested in the small intestine.

        ✓ Used directly by the animal because it is absorbed as small proteins and amino acids.

    o **Metabolizable Protein** (MP) is protein available to the animal including microbial protein (BCP) synthesized by the rumen microorganisms and UIP.

    o **Digestible Protein** (DP) is reported by some laboratories but protein digestibility is influenced by external factors.

- Fiber is determined as crude fiber, neutral detergent fiber or acid detergent fiber.

    o **Crude Fiber** (CF): traditional measure of fiber content in feeds.

- ✓ Neutral detergent fiber (NDF) and acid detergent fiber (ADF) are more useful measures of feeding value and should be used to evaluate forages and formulate rations.

  - **Neutral Detergent Fiber** (NDF): Structural components of the plant, specifically cell wall. NDF is a predictor of voluntary intake because it provides bulk or fill.

    - ✓ In general, low NDF values are desired because NDF increases as forages mature.

  - **Acid Detergent Fiber** (ADF): Least digestible plant components, including cellulose and lignin.

    - ✓ ADF values are inversely related to digestibility, so forages with low ADF concentrations are usually higher in energy.

- Fat is sometimes referred to as the ether extract (EE) portion of a feedstuff.

  - Provides an estimate of crude fat content of a feedstuff.

  - Fat is an energy source with 2.25 times the energy density of carbohydrates and proteins.

- **Nitrogen-Free Extract** (NFE) represents carbohydrates, sugars, starches and a major portion of materials classed as hemicellulose in feeds.

  Chapter Resource

  - When crude protein, fat, water, ash and fiber are added and the sum is subtracted from 100, the difference is NFE.

Energy Terms and Calorie

- **Total Digestible Nutrients** (TDN): sum of the digestible fiber, protein, lipid and carbohydrate components of a feedstuff or diet.

  - Directly related to digestible energy and is often calculated based on ADF

  - Useful for beef cow rations that are primarily forage

  - Tends to under predict the feeding value of concentrate relative to forage

- **Net Energy** (NE): Mainly referred to as net energy for maintenance (NEm), net energy for gain (NEg) and net energy for lactation (NEl).

  - **Net energy** system separates the energy requirements into their fractional components used for tissue maintenance, tissue gain and lactation.

- Expressed as megacalories per pound (Mcal/lb).
- Accurate use of the NE system relies on careful prediction of feed intake.
- In general, NEg overestimates the energy value of concentrates relative to roughages.

• A **calorie** is the basic unit of energy. Defined as the amount of energy needed to raise one gram of water one degree Celsius, measured from 14.5°C to 15.5°C.

- The calorie is used to measure and identify the total amount of energy in a feed.

Feed Classifications

• Classes of feeds are divided into and defined as:

- **Concentrates** – Feeds that are low in fiber and high in total digestible nutrients; for example, grains or oilseed meals.
- **Roughages** or **Forages** – Feeds that are high in fiber and low in total digestible nutrients; for example, hay or grass.

Chapter Resource

Importance of Feed Composition

• Every animal has specific needs for feed.

• This need varies with age, activity level, climate and stage in the production cycle.

• A knowledge of the amount of the nutrients needed by the body and the value of the feed as a source of the nutrients is the basis for good management and proper diets.

• Feed composition information gained slowly through research and trial and error.

Summary

Feeds are classified according to chemical analysis into six general groups: water/moisture, mineral matter or ash, protein, fiber, fat and nitrogen-free extract. Water is expressed as the percentage of moisture in a feed. Mineral/ash measures the total mineral content in a feed but not individual minerals. Protein contains the nitrogen (N) in the feed and early measurements of protein content were determined by the amount of N in the feed sample. Fiber is determined as crude fiber, neutral detergent fiber or acid detergent fiber. Fat is sometimes referred to as the ether extract (EE) portion of a feedstuff. Total Digestible Nutrients (TDN): sum of the digestible fiber, protein, lipid and carbohydrate components of a feedstuff or diet. Net Energy (NE): Mainly referred to as net energy for maintenance (NEm), net energy for gain (NEg) and net energy for lactation (NEl). Classes of feeds are divided into: Concentrates and roughages or forages. Importance of Feed Composition: Every animal has specific needs for feed. Knowledge of nutrients and amounts is necessary for good management.

Additional Resources

Pond, W.G., D.C. Church, K.R. Pond, P.A. Schoknecht. 2004. Basic animal nutrition and feeding. 5th ed. Hoboken, NJ: Wiley & Sons Inc.

Salem, A.F.Z.M. 2012. Nutritional strategies of animal feed additives. Nova Science Pub, Inc.

Feed Additives for Beef Cattle
http://www.uaex.edu/Other_Areas/publications/pdf/FSA-3012.pdf

Feedstuffs for Beef Cattle
http://extension.msstate.edu/publications/publications/feedstuffs-for-beef-cattle

Swine Feed Additives
http://www.ncsu.edu/project/swine_extension/nutrition/nutritionguide/feed%20additives/feedadd.htm

Assessment

Take assessment online here: http://tinyurl.com/AnSci-FeedComposition
Download and print the assessment by scanning this QR code or by going to this URL: http://www.tagmydoc.com/Ch32AnSci

33 Feed Preparation

Major Concept
Feed preparation affects palatability, use and storage.

Objectives
- Identify four methods of feed preparation with associated advantages and disadvantages
- Name a fermented feed

Key Terms
- Anaerobic
- Fermenting
- Pellets

Chapter Resource

 Complementary *full color* illustrations, photos, charts and graphs are available by scanning this QR code or by following this URL: http://www.tagmydoc.com/AS33 This digital resource will enhance your understanding of the chapter concepts.

Feed Preparation Methods
- Profitability of grinding, crushing, soaking feed or **fermenting** (a process that converts sugar to acids, gases and/or alcohol).
  - Only when animals fail to chew the grain thoroughly and thus lose the nutritive value of the feed because of the reduction of surface area.

- Grinding grain
  - Ground to medium-fine
  - Extremely fine grinding takes more power and time; the process makes grain less palatable.
  - Grain too fine could be wasted and/or cause digestive problems.

JTF Introduction to Animal Science 193

- Rolling vs. grinding grain

  o Rolling is more palatable, higher in bulk and may be less likely to cause digestive disturbances in heavy feeding.

- Some feeds are made into **pellets** (a small, condensed formed feed) which have advantages:

  o Less waste from wind and weather.

  o Less space needed for storage.

  o Animals may eat more feed in pelleted form.

Chapter Resource

- Chopped or ground hay

  o Advisable only if it encourages animals to eat coarse, stemmy portions that might otherwise be left behind.

- Feeding 100% concentrate

  o Not advisable because of digestive troubles in animals, especially ruminants.

- Cooked feed

  o Advantageous only when feeding potatoes, field beans and soybeans.

  o Cooking does not increase the digestibility or feeding value of other feeds.

- Soaking feeds

  o Advantageous only when grain with small or hard kernels cannot be ground or crushed.

  o If soaked too long, grain will become stale and fermentation may take place.

- Fermenting of feeds

  o Common types of silage:

    ✓ Haylage
    ✓ Corn silage

Corn Silage

  o Harvesting corn as silage furnishes 50-60% more nutrients per acre for beef cattle than harvesting the grain alone.

- Process must be carried out under acidic conditions (around pH 4-5) in order to keep nutrients and provide a form of food that cows and sheep will like to eat.

- Fermentation at higher pH results in silage that has a bad taste and lower amounts of sugars and proteins.

- Removing and keeping out oxygen is a key part of making silage. This is because fermentation occurs under **anaerobic** (oxygen-free) conditions, or the correct type of microorganisms won't grow.

Summary

Extremely fine grinding takes more power and time and the process makes grain less palatable. Rolling is more palatable, higher in bulk, and may be less likely to cause digestive disturbances in heavy feeding. Advantage of pellets: less waste from wind and weather, less space needed for storage and animals may eat more feed in pelleted form. Feeding 100% concentrate is not advisable because of digestive troubles in animals, especially ruminants. Cooking does not increase the digestibility or feeding value of other feeds. Soaking feeds is advantageous only when grain with small or hard kernels cannot be ground or crushed. Corn and hay are often fermented to produce silage.

Additional Resources

Kellems, R.O. and D.C. Church. 2009. Livestock feeds and feeding. 6th ed. Upper Saddle River, NJ: Prentice Hall.

Morrison, F.B., E.B. Morrison, S.H. Morrison and R.B. Morrison. 1951. Feeds and feeding: A handbook for the student and stockman. 21st ed. Unabridged. 50th Anniversary. Morrison Publishing Co.

Corn Silage for Beef Cattle
http://extension.missouri.edu/p/G2061

Feeding Beef Cattle
http://extension.psu.edu/business/ag-alternatives/livestock/beef-and-dairy-cattle/feeding-beef-cattle

Loy, D. 2015. Ten ways to cut cattle feeding costs.
https://www.extension.iastate.edu/agdm/livestock/html/b1-71.html

Assessment

 Take assessment online here: http://tinyurl.com/AnSci-FeedPreparation
Download and print the assessment by scanning this QR code or by going to this URL: http://www.tagmydoc.com/Ch33AnSci

34 Feed Additives

Major Concept

To enhance livestock production, producers use feed additives.

Objectives

- Identify the functions of feed additives
- Name two common feed additives

Key Terms

- Feed additive
- Food nutrient
- Surfactants

Chapter Resource

Complementary *full color* illustrations, photos, charts and graphs are available by scanning this QR code or by following this URL: http://www.tagmydoc.com/AS34 This digital resource will enhance your understanding of the chapter concepts.

Feed Additives Defined

- Feed Additive vs. Food Nutrient

    o A **feed additive** does not become a part of the animal's body.

    o A **food nutrient** becomes a part of the body cells and is necessary for the proper function of these cells.

- Functions of feed additives:

    o Promote growth and production

    o Improve feed efficiency

    o Provide prevention and treatment for internal parasites and diseases

- - Provide specific actions or results in specific species (e.g., what works on pigs may not work on cattle)
- Common feed additives include: antibiotics, hormones, arsenicals, bacteria and surfactants.
  - Antibiotics do not cause a marked change in meat quality; any effects would be indirect, resulting from improved general health (more rapid weight gains).
    - ✓ Examples: Areoumycin; Terramycin; Penicillin; Zinc Bacitracin; Hygromycin B; and Streptomycin
    - ✓ While the uses of the antibiotics vary, in many cases use of them promotes growth (INDIRECTLY) and reduces the number of "unthrifty" animals.
  - Hormones (see Hormones in Meat Production)
  - Dry rumen bacteria can be fed to cattle to stimulate rumen bacterial development.
  - Detergents or **surfactants** (ingredients that reduce surface tensions of liquids) are used to reduce and stop foaming to prevent bloat.

Feed Additives

- Erythromycin (improves growth) – antibiotic
- Neomycin (prevents scours) – antibiotic
- Ethylene diamine dihydroiodide is used to prevent foot rot and lumpy jaw.
- Poloxalene is a surfactant used to control or prevent bloat.
- Ammonia chloride is a chemical used to prevent urinary calculi.
- Ronnel is for control of grubs.
- Thibenzole controls internal parasites.
- Melengestrol Acetate (MGA) is used to depress heat in heifers which generally improves feedlot performance.

Chapter Resource

Feed Additives Sources

- Commercially prepared feeds may contain additives.

- Many, if not all, additives must be stopped at a predetermined time BEFORE slaughter. This gives the animal's time to "clear" any residues which could be harmful to consumers.

Summary

Feed additives function to promote growth and production, improve feed efficiency, provide prevention and treatment for internal parasites and diseases and provide specific actions or results in specific species. Common feed additives are antibiotics, hormones, arsenicals, dry rumen bacteria, detergents or surfactants and other additives. Commercial prepared feeds (no farm-produced feeds contain additives) contain additives.

Additional Resources

Pond, W.G., D.C. Church, K.R. Pond, P.A. Schoknecht. 2004. Basic animal nutrition and feeding. 5th ed. Hoboken, NJ: Wiley & Sons Inc.

Salem, A.F.Z.M. 2012. Nutritional strategies of animal feed additives. Hauppauge, NY: Nova Science Pub, Inc.

Feedstuffs for Beef Cattle
http://extension.msstate.edu/publications/publications/feedstuffs-for-beef-cattle

Medicated Feed Additives for Cow-Calf and Stocker/Backgrounding Production Systems
http://www.uaex.edu/publications/pdf/FSA-3012.pdf

Swine Feed Additives
http://www.ncsu.edu/project/swine_extension/nutrition/nutritionguide/feed%20additives/feedadd.htm

Assessment

Take assessment online here: http://tinyurl.com/AnSci-FeedAdditives
Download and print the assessment by scanning this QR code or by going to this URL: http://www.tagmydoc.com/Ch34AnSci

35 Computing Balanced and Least Cost Rations

Major Concept

Efficient livestock production requires that animals be fed a balanced ration.

Objectives

- Identify the components of a simple feeding regime for market swine, lambs and beef
- Name three methods for balancing a ration
- Distinguish between a balanced ration and an economically balanced ration for livestock
- Identify the steps in developing a least-cost ration
- List three general guidelines for feeding livestock

Key Terms

- Balanced ration
- Pearson square
- Ration

Chapter Resource

 Complementary *full color* illustrations, photos, charts and graphs are available by scanning this QR code or by following this URL: http://www.tagmydoc.com/AS35 This digital resource will enhance your understanding of the chapter concepts.

Livestock Rations

- Balancing a ration for livestock

- **Ration** is:

    o Feed allowed for a given animal during a 24-hour period, whether it is fed at one time or in portions at different times during the day.

- **Balanced ration**:

- o Contains more than one feed, so proper quantities of essential nutrients will be provided.

- o Furnishes the required nutrient in such proportion and amount that the animal is properly nourished for the 24-hour period.

- Three methods for balancing rations:

    Chapter Resource

    1. Square method (used for calculation of protein since protein is often a limiting factor in a ration).
    2. Feeding standard (used for all animal requirements).
    3. Computer formulated rations (least-cost).

- o Feeding standards are found in tables stating the amounts of nutrients which should be provided in rations for specific animals of various ages and classes to secure the best results (found by themselves and/or in nutrition and animal science texts; also published by the NRC [National Research Council]).

General Guidelines

- Identify animals to be fed by age, weight, function and specific conditions under which they are fed.

- Select nutrient requirements and allowances.

- Make ration palatable, safe, economical and nutritionally balanced.

- Determine fixed amount of vitamins and minerals then mix grains relative to protein supplement for desired protein level.

Square Method (called Pearson Square):

- Pearson square helps to formulate seed rations.

- Step 1:

    o Determine (using the appropriate table) the percent protein needed.

    o Decide what protein supplement(s) and which grain(s) will be used in the ration.

- Step 2:

- Construct a square in the middle of the paper.
- Put the desired crude protein percentage of the animal's feed in the middle of the square.
- Put the crude protein percentages of the feed on the left-hand side of the square.
- Supplement (may be roughage) in lower corner.
- Grains (or average of grains) in upper corner.
- Subtract diagonally and place the differences at the right-hand corners.
- Upper right gives parts of supplements.
- Lower right gives parts of grain.
- Add the values on the right side of the square and place that sum below each of the numbers in the right corners; this represents (in fraction form) the portions of each feed needed to mix the ration balanced for protein.

Chapter Resource

- A **Pearson Square** completed:

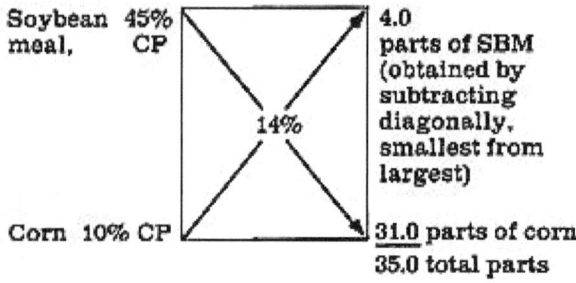

Soybean meal, 45% CP

Corn 10% CP

14%

4.0 parts of SBM (obtained by subtracting diagonally, smallest from largest)

31.0 parts of corn
35.0 total parts

Feeding Standards Method

- Method requires using a worksheet (sample below) for the specific type of livestock being fed.
- A trial-and-error method:
- Step 1:
  - Producer determines the type and class of animal being fed.
- Step 2

- Then producer decides which feeds will be used, depending on availability and cost.

- Step 3:

  - Referring to feed composition tables, the producer finds the nutrient content of the selected feeds, the requirements of total protein, TDN, digestible energy, or net energy, calcium and total pounds required/day for the animal.

  - These values are added to the worksheet.

Chapter Resource

- Step 4:

  - Producers then estimate the amount of each ingredient to feed the animal remembering that the total for the ingredients cannot be larger than the requirement.

    - ✓ Number of pounds added to the worksheet. This is the trial-and-error part of the problem.

- Step 5:

  - Pounds of individual feed are then multiplied times the percentages of feed values for each of the nutrient categories

- Step 6:

  - Result are added and observed if the totals are even or above the animal's requirements.

  - If necessary the individual pounds of feed are adjusted and re-multiplied and get as close as possible to the requirements of the animal.

- Sample of worksheets used to balance a ration:

Least Cost Balanced Rations

- Definition: A ration that fulfills all nutritional requirements of a particular group of animals but uses the mix of ingredients from the list of those available that costs the least.

- Uses a computer program

- Relies of the substitution of feeds based on cost and nutritional value

- Producers make changes and compare the cost of a new ration with an old one and determine if the substitution pays or not.

- Rules of Thumb:

    o Silage can be substituted for hay at a rate of 3 pounds of silage for 1 pound of hay.

    o Feeds which can be substituted for alfalfa:

    - ✓ Clover
    - ✓ Lespedeza
    - ✓ Timothy hay
    - ✓ Prairie hay

    o Feeds which can be substituted for corn in rations:

    - ✓ Barley
    - ✓ Milo
    - ✓ Sorghum
    - ✓ Oats
    - ✓ Rye
    - ✓ Wheat

    o Protein supplements substitutions:

    - ✓ Linseed meal, soybean meal and cottonseed meal may be used almost interchangeably depending on which is more economical
    - ✓ Tankage and meat scraps
    - ✓ Soybeans
    - ✓ Wheat bran

    Chapter Resource

    o Least cost balanced rations should include the following factors:

    - ✓ Use feeds the producer has available at the lowest cost.
    - ✓ Meet minimum requirements from the feeding chart(s).
    - ✓ Use proper proportions of concentrates and roughages in the ration.

Guides for Concentrates and Roughages

- Swine:

    o Fattening pigs in drylot: 5-15% legume hay in ration.

    o Pregnant sows in drylot: 10-15% legume hay in ration.

    o Pregnant sow: 1-1.5 lbs of concentrate per 100 lbs body weight.

    o Pregnant gilts: 1.5-2.0 lbs of concentrate per 100 lbs body weight.

    o Fattening swine will consume 5 lbs of feed (mostly concentrates) per 100 lbs of body weight.

Chapter Resource

- Beef Cattle:

    o Fattening cattle: 2.1 lbs or more of concentrates and dry roughages daily per 100 lbs of body weight. Allowance on concentrates ranges from less than 1 lb. to 1.7 lbs or more daily per 100 lbs of live weight.

    o Full feed: Cattle will consume 1 3/4 - 2 1/2 lbs of concentrates and 3/4 - 2 lbs of hay per 100 pounds of live weight daily.

    o Breeding cows: If roughage is low in protein, 1 lb. per head daily of protein supplement should be fed along with a small amount of grain.

- Dairy Cattle:

    o Dairy cows in milk: 2.0 lbs of dry roughage daily per 100 lbs of live weight. For concentrates use the tables in a feeds reference book.

    o Dairy heifers: Up to 6 months - 2 to 3 lbs of concentrates per head per day on good roughage. 4 to 5 lbs. on poor roughage. Over a year old: 2 - 4 lbs of concentrates on fair roughage.

- Sheep:

    o Breeding ewes: Good roughage up to 4-6 weeks before lambing. During 4-6 weeks, 0.5 to .75 lbs of concentrates per head per day.

    o Nursing ewes: Ewes not on pasture need 1 lb. or more per head per day of concentrates in addition to good roughage.

    o Fattening lambs: 70 lbs, 1.5 lbs. of roughage per head per day and all the grain they will eat. 2.5 lbs or more of hay per head per day when grain is restricted.

Full feed: lambs will consume about 2 lbs of grain and about 2 lbs of roughage per head per day.

Computing the Cost of a Ration

- Figure cost of feed to price per pound.

    o Take a one-ton price and divide by 2000 to get the price per lb.

    o Multiply this amount by the lbs of feed used per day in the ration.

    o Add together the figures for all the feeds used and divide by the numbers of animals fed to get the cost of feed per animal per day.

The Art and Science of Feeding

- Livestock feeding is as much art/observation as it is science due to individual variability of livestock.

- Sudden feed changes:

    o Digestive disturbances will result and animals will go off feed; in certain instances, they may even die.

Chapter Resource

- Least-cost balanced ration vs. balanced ration:

    o A ration will not be economical unless the animals like it, it keeps them in good health and they grow, fatten or produce well on it.

    o A balanced ration is not necessarily the most economical ration to feed livestock.

Summary

A ration is the feed allowed for a given animal during a 24-hour period, whether it is fed at one time or in portions at different times during the day. A balanced ration generally contains more than one feed, so proper quantities of essential nutrients will be provided and is one which furnishes the required nutrient in such proportion and amount that the animal is properly nourished for the 24-hour period. There are three methods for balancing rations: the square method, the feeding standard and computer formulated rations. Rule of thumb: Silage can be substituted for hay at a rate of 3 pounds of silage for 1 pound of hay. Substitutions for alfalfa are clover, lespedeza, timothy hay and prairie hay. Substitutions for corn in rations include barley, milo, sorghum, oats, rye and wheat. Protein supplements include linseed meal, soybean meal and cottonseed meal.

A ration will not be economical unless the animals like it, it keeps them in good health, and they grow, fatten or produce well on it.

Additional Resources

Parker, R. 2008. Equine science. 3rd ed. Clifton Park, NY: Thomson Delmar Learning. (Pg. 248-262).

Taylor, R.E. and T.G. Field. 2010. Scientific farm animal production: Introduction to animal science. 10th ed. Upper Saddle River, NJ: Prentice Hall.

Formulating rations with the Pearson Square
http://www.ext.colostate.edu/pubs/livestk/01618.html

Lamb Feedlot Nutrition
http://extension.colostate.edu/topic-areas/agriculture/lamb-feedlot-nutrition-1-613/

Ration balancing, by hand or by computer,
http://www.ansc.purdue.edu/compute/balance.htm

Assessment

Take assessment online here: http://tinyurl.com/AnSci-ComputingRations
Download and print the assessment by scanning this QR code or by going to this URL: http://www.tagmydoc.com/Ch35AnSci

36 Lactation

Major Concept

Efficient milk production requires an understanding of the process of lactation.

Objectives

- Identify hormones that control the growth and development of the mammary gland and the production of milk
- Name parts associated with the structure of the mammary gland
- Identify a lactation curve
- List factors that affect milk production
- Define colostrum and identify its importance to the young

Key Terms

- Alveoli
- Colostrum
- Epinephrine
- Exocrine gland
- Lactation
- Lactation curve
- Lobules
- Mastitis
- Passive immunity

Chapter Resource

 Complementary *full color* illustrations, photos, charts and graphs are available by scanning this QR code or by following this URL: http://www.tagmydoc.com/AS36 This digital resource will enhance your understanding of the chapter concepts.

Lactation Defined

- Production of milk by/from the mammary gland.

- Milk provides nutrition and **passive immunity** (acquired by transferring of antibodies from an immunized animal to an unimmunized one) to young.

    o Controlled by hormones

    o Source of nutritional products for humans

Mammary Gland Structure

- Support
    - By ligaments
- Structural components
    - **Alveoli** (hollow cavities)
    - Milk duct
    - Gland cistern
    - Annular ring
    - Teat cistern
    - Streak canal

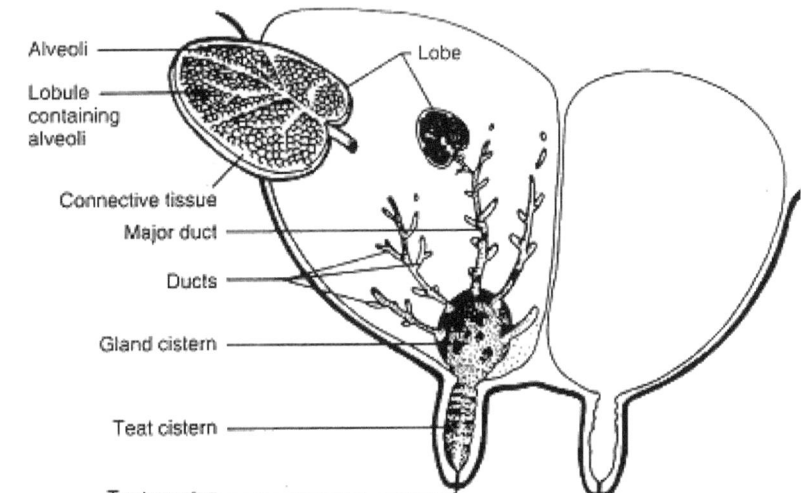

- Considered an **exocrine gland** (secrete their products into ducts) – produces an external secretion (milk) from its alveoli.

Species Differences

- Cow
    - 4 quarters
- Ewe, Doe, Mare
    - 2 halves
- Sow
    - 8 to 20 mammary glands

Mammary Gland Development and Milk Secretion

- Hormones of hypothalamus, pituitary, thyroid, adrenal, ovary and placenta involved:
    - From the pituitary: ACTH, TSH, FSH, LH, Growth Hormone, Prolactin and Oxytocin.

- From the thyroid: Thyroxine (T4).
- From the ovaries and placenta: Estrogen and progesterone.
- Hormone epinephrine suppresses lactation/milk let-down.

Maintenance of Lactation

- Milking or the act of lactating causes it to be maintained.
- Reproduction: renewed each time the animal gives birth.

Factors Affecting Milk Production

- **Mastitis** (inflammation of the mammary gland in the udder) decreases milk production.
- Demand increases
- Male birth increases
- Multiple births increase

Chapter Resource

Nutrients

- Provides for the early growth of young animal
- Milk Composition (Percentage)

Species	Solids	Fat	Protein	Lactose
Cow	12.7	3.9	3.3	4.8
Sow	19.0	6.8	6.3	5.0
Goat	12.4	3.7	3.3	4.7
Human	13.3	4.5	1.6	7.0

Using Dairy Cow as an Example

- To produce milk, a cow needs to have a calf.
- After birth, milk production peaks and then gradually decreases.
- Birth of a calf stimulates hormone production which causes milk letdown.

- After approximately 305 days, if not milked, the cow should "go dry", or stop producing milk.

- If cow is not re-bred, she will not produce any more milk.

- An approximate 60-day drying off period is vital to milk production because it allows time for the udder to heal.

- Milking, either by hand or mechanically, stimulates milk production over the lactation period.

- Lactation period is best shown by **lactation curve** (period during which the mammary glands secrete milk).

- Other livestock have a lactation curve unique to each.

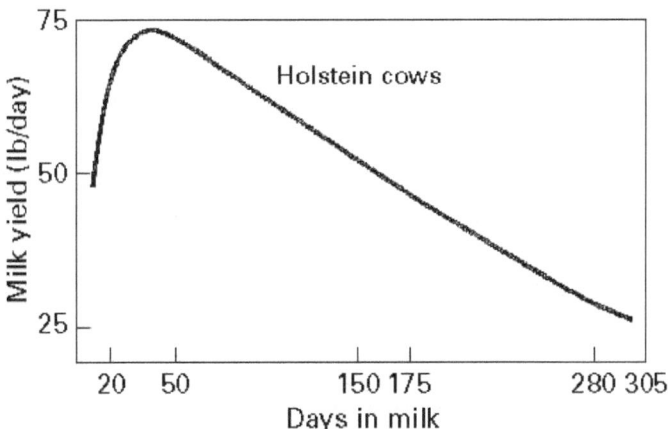

Importance of Feeding Colostrum to Calves

- **Colostrum** is the first milk to come from a cow/sow/ewe/mare after birth and contains a high concentration of antibodies, providing passive immunity to disease through immunoglobulins (Ig)

- Young animal's intestines at birth are very porous, which allows it to absorb the antibodies from the milk.

- Intestines begin to "close up" from 24-48 hours after birth, so it is imperative that the young suckles within the first 24 hours of life.

- Calf/piglet/lamb/foal must ingest the colostrum in the first 24 hours because colostrum gives it immunity against disease.

Tracing Milk from the Udder to the Milker

- Prolactin (from the pituitary) stimulates alveoli cells to produce milk.

- Milk drains into the lumen of the alveoli.

- Clusters of alveoli, called **lobules**, contain ducts that drain into larger ducts. These larger ducts drain into the gland cistern.

- Milk is stored in the gland cistern.

- Sphincter muscles prevent milk from leaking into the teat.

Chapter Resource

Milk Letdown

- Hormone oxytocin is released by the pituitary gland into the bloodstream when the udder is stimulated.

- Oxytocin causes the alveoli to "squeeze" and release the milk.

- Oxytocin release can be caused by: washing the udder prior to milking, suckling of the calf/young animal, or other pleasant stimuli.

- If an animal becomes frightened or upset, a hormone called **epinephrine** (adrenaline) is released that inhibits milk letdown.

- Milking time must be a non-frightening experience for the cows every time.

Summary

Lactation is the production of milk from the mammary gland. It is controlled by hormones. Milk provides nutrition and passive immunity to young through immunoglobulins. Mammary glands are supported by ligaments and structural components. Maintenance of lactation involves the act of lactation and reproduction. Factors affecting milk production are mastitis, demand, male birth and multiple births. Colostrum is produced for only a short time following birth. Absorption time is critical for the young and must occur within the first hours of birth. The hormone oxytocin causes milk letdown. If an animal becomes frightened or upset, a hormone called epinephrine is released that inhibits milk letdown.

Additional Resources

Taylor, R.E. and T.G. Field. 2011. Scientific farm animal production. 10th ed. Upper Saddle River, NJ: Prentice Hall.

Colostrum Supplements and Replacer
http://extension.psu.edu/animals/dairy/health/nutrition/calves/colostrum/das-11-180

Lifecycle Production Phases
http://www.epa.gov/oecaagct/ag101/dairyphases.html

Mammary Macro-Structure Dairy Cow Udder Anatomy
http://ansci.illinois.edu/static/ansc438/Mamstructure/anatomy_1.html

Assessment

 Take assessment online here: http://tinyurl.com/AnSci-Lactation
Download and print the assessment by scanning this QR code or by going to this URL: http://www.tagmydoc.com/Ch36AnSci

37 Factors Affecting Animal Health

Major Concept

Since many factors affect animal health, preventative herd health management programs help to eliminate or reduce these factors.

Objectives

- Identify the seven major components of an animal health program and their role in maintaining health
- List four stresses that could possibly impair an animal's immune system

Key Terms

- Antiseptics
- Disinfectants
- Sanitation
- Stress

Chapter Resource

 Complementary *full color* illustrations, photos, charts and graphs are available by scanning this QR code or by following this URL: http://www.tagmydoc.com/AS37 This digital resource will enhance your understanding of the chapter concepts.

Preventative Herd Health Management Programs

- Help to eliminate or reduce most of the health-related livestock losses.

- Most major animal disease problems are associated with health management.

- Components of a herd health management program include:

    o Veterinarian-Assisted Planning provides the producer with professional advice concerning vaccinations, cost effectiveness of treatment and other preventative medical practices.

    o **Sanitation** or cleanliness: The severity of some diseases is dependent on the number and virulence of microorganisms entering the animal's body.

    o Many organisms live and multiply outside the animal, so the number of microorganisms can be reduced by implementing sanitation practices.

- ✓ Microorganisms thrive in manure and other organic waste materials, therefore good sanitation practices must be in place. Buildings, pens, and pastures should be well drained, preventing prolonged wet areas or mud holes.

- ✓ Antiseptics and disinfectants can be effectively utilized in a good sanitation program. **Antiseptics** are substances, usually applied to animal tissue, that kill or prevent the growth of microorganisms. **Disinfectants** are products that destroy pathogenic microorganisms. They are agents used on inanimate objects. In absence of disinfectants, sanitizing with clean water is helpful.

o Proper nutrition: Well-nourished animals receive an adequate daily supply of essential nutrients.

- ✓ Undernourished animals usually have a weak immune system, thus making them more vulnerable to invading microorganism.

o Physical facilities contribute to animal health problems by causing physical injury or stress, or by allowing spreading of pathogens through a group of animals.

- ✓ They can contribute to the spread of disease by not preventing its transmission (e.g. venereal disease transmission owing to poor or inadequate fences).
- ✓ Even proper facilities that are misused, such as feedlot, can provide an easier transmission of disease owing to crowding, stress and poor sanitation.

o Proper use of biologics and pharmaceuticals:

Chapter Resource

- ✓ Biologics are used primarily to prevent diseases by stimulating the animal's immune system against specific diseases and others stimulate the body to produce antibodies that fight disease.
- ✓ Pharmaceuticals are used to kill or reduce the growth of microorganisms in the treatment of diseases and infections.
- ✓ Examples include antibiotics, steroids, sulfa compounds and hormones.

o Minimizing stress: **Stress** is any environmental factor that can cause a significant change in the animal's physiological processes. Prolonged stress can impair the body's immune system, causing a reduced resistance to disease.

o Examples of physical sources of stress include:

- ✓ Weather (extreme temperatures, wind, snow, mud, dust)
- ✓ Nutritional deficiency or oversupply
- ✓ Fatigue
- ✓ Weaning

- ✓ Transportation
- ✓ Castration
- ✓ Dehorning
- ✓ Abusive handling
- ✓ Over crowding

Chapter Resource

o Record keeping: Proper records permit health problems to be identified, determine what is causing them and allowing alternative methods of prevention and treatment to be assessed.

- ✓ Many software programs are available to manage records.

Summary

Preventative herd health management programs can eliminate or reduce most of the health-related livestock losses. Most major animal disease problems are associated with health management. The components of a herd health management program include: veterinarian-assisted planning, sanitation, proper nutrition, physical facilities, proper use of biologics and pharmaceuticals, minimizing stress and record keeping.

Additional Resources

Beef Cow/Calf Herd Health Program and Calendar
http://pubs.ext.vt.edu/400/400-007/400-007_pdf.pdf

Herd-Health Programs for Limited-Resource Farmers: Prevention vs Treatment
http://www.joe.org/joe/2011october/comm2.php

Herd Health Programs for Swine Seedstock Production
http://extension.missouri.edu/p/G2508

Herd Management for Disease Prevention
http://extension.missouri.edu/p/G2507

The U.S. Pork Industry's Herd Health Practices
http://www.uslge.org/porkherd_health_.html

Beef Herd Health
http://extension.psu.edu/animals/beef/nutrition/articles/beef-herd-health

Assessment

 Take assessment online here: http://tinyurl.com/AnSci-Factors
Download and print the assessment by scanning this QR code or by going to this URL: http://www.tagmydoc.com/Ch37AnSci

38 Animal Health Evaluation

Major Concept

Animal health evaluations require one to identify an unhealthy animal through observations and the measurement of body temperature, pulse and respiration rate.

Objectives

- Identify visual and non-visual indicators of unhealthy animals
- Recognize normal body temperature, pulse rate and respiration rate of various livestock

Key Terms

- Dehydrated
- Demeanor
- Mucous membranes
- Turbidity

Chapter Resource

 Complementary *full color* illustrations, photos, charts and graphs are available by scanning this QR code or by following this URL: http://www.tagmydoc.com/AS38 This digital resource will enhance your understanding of the chapter concepts.

Animal Health Observed

- A producer is the most vital link to the continued and progressive health of his/her animals and can also interpret symptoms of illness and convey those to a health professional.

- To interpret irregular behavior as an illness or lack of health, one must first understand the normal behavior of that animal or species of animal.

  o Normal behavior is determined by time spent observing the habits of a normal and healthy animal.

  o Once normal behavior is determined, abnormal behavior is easier to identify by contrast.

  o What may be a normal habit for one animal or group may be abnormal for another.

Physical Examinations

- Involves using all five senses: listening, touching, seeing, smelling and, at times, tasting.

  o Inspection at a distance should be done first, look at the animals:

    ✓ Sleeping habits – time of day, standing, or lying
    ✓ Eating habits/nutritional state – time spent eating, time of day preferring to eat, meal eaten throughout day or all at once
    ✓ Attitude – gregarious, pecking order, social behavior, solitude
    ✓ Posture – weight shifting from one foot or side
    ✓ Gait – is there indication of lameness
    ✓ Breathing – difficult, fast, slow
    ✓ Physical condition – weight loss, hair coat, skin odor

    Chapter Resource

  o Next, restrain the animal and note:

    ✓ Temperature
    ✓ Heart rate (pulse)
    ✓ Respiratory rate
    ✓ Restraint method could cause these to increase.

  o Then, begin a systematic examination of all major areas of the animal body, noting amount, color, **turbidity** (muddiness created by stirring up sediment or having foreign particles suspended), odor and consistency of discharges from any of the following body openings:

    ✓ Nose
    ✓ Mouth
    ✓ Eyes
    ✓ Ears
    ✓ Uro-genital (urinary, genital)
    ✓ Anus – take samples of feces and urine if needed.
    ✓ Any wounds, scratches or abrasions.

  o Continue to review external body surfaces; look for changes in color, size and shape indicating abnormalities.

  o Evaluate the:

- ✓ Status of hydration – does the animal look **dehydrated** (body is lacking water).

o Color of **mucous membranes** (an epithelial tissue that secretes mucus which lines many body cavities and tubular organs including the gut and respiratory passages) – which should be pink, not yellow or white.

- ✓ Capillary refill time – should be quick.
- ✓ Mouth/throat – too much saliva may indicate infection in mouth or inability to swallow, lack of saliva may indicate fever or colic, check for choking, coughing, drooling, gagging, vomiting, foul odor, difficulty swallowing.
- ✓ Ears – no discharge, head should not be tilted
- ✓ Ribs – labored breathing
- ✓ Legs/feet – lameness, stiffness, straining, any leg favored
- ✓ Genitals – swelling or rash
- ✓ Hair coat/skin – parasites, abnormal roughness
- ✓ Swelling/lumps/wounds – wound discharge, ulcerations
- ✓ Position of the animal – is it down, unable to rise, uncoordinated
- ✓ Temperature, does it have a fever – dry muzzle, loss of appetite, thirst, constipation, indigestion
- ✓ Milk production – normal for this animal
- ✓ Possibility of the animal being in pain – are there tender spots, or things the animal avoids doing that are abnormal?

Chapter Resource

o After a thorough review, the symptoms and clinical signs will serve to narrow and determine the source of the problem.

o An important symptom of any animal is pain and it should always be a major concern when evaluating an animal's overall health.

o Pain can be indicated by an animal in many ways, including:

- ✓ Irritability
- ✓ Avoidance of certain activities
- ✓ Atypical fear

o **Demeanor** (outward behavior) – not alert, hunched up, etc.

- ✓ Loss of appetite

- Producers know the normal range of the temperature for the species before determining if the animal's temperature is abnormal.

    o Chart below summarizes normal temperature ranges.

    o Air temperature, exercise, late stages of pregnancy and illness may affect normal temperatures.

    o Rectal Temperatures

Animal	F (+ or -) 1 degree	C (+ or -) 0.5 degree
Cat	101.5	38.5
Cattle	101.5	38.5
Dog	102	39
Goat	104	40
Horse	100.5	38
Pig	102	39
Rabbit	102.5	39.3
Sheep	103	39.5

- Heart Rates (Beats/Minute)

Animal	Average	Range
Cat	120	110 – 140
Chicken	--	250 – 300
Cow	--	60 – 70
Dog	--	100 – 130
Goat	90	70 – 135
Horse	44	28 – 70
Human	70	58 – 104
Rabbit	205	128 – 304
Sheep	75	60 – 120
Swine	--	59 – 86

Chapter Resource

- Determining Pulse Rate

    o Take pulse for 15 seconds and multiply by 4 to get pulse rate per minute.

    o Best method includes the use of a stethoscope, but if none is available, pulse rate can be determined by constricting a major artery, or even listening with an ear, or feeling the heart beat through the ribs with a hand.

    o For the following animals, the best place to get a pulse is:

    ✓ Cattle – press the artery outside of jaw just above the lower border, use the soft spot just above the inner dewclaw (a vestigial digit on the foot of many mammals), or just above the hock (the joint in the hind leg between the knee and the fetlock).

- ✓ Sheep and swine – press the artery inside the thigh where it comes close to the skin surface.

- ✓ Horses – press the artery of jaw just in front of throatlatch, artery inside of elbow, the artery under tail, or inside of the thigh.

- ✓ Lab animals, dogs and cats – with smaller animals, hold the chest in palm of hand with fingers around the rib cage and feel heartbeat, with larger animals, press the artery inside thigh.

- Determining Respiration Rate:

  o Use a stethoscope, if available.

  o With small animals, count the number of times the ribs rise and fall.

    ✓ With large animals, place hand on the flank (side of an animal's body between the ribs and the hip) and count the rise and fall of the abdomen for one minute.

  o Respiratory Rate (breaths/minute)

Chapter Resource

Animal	Respiration Rate
Cat	20-30
Chicken	12-36
Cow	8-16
Dog	10-30
Goat	12-15
Horse	8-16
Human	12-18
Pig	15-30
Rabbit	30-60
Sheep	12-20

Summary

Normal behavior is determined by time spent observing the habits of a normal and healthy animal, realizing that what may be a normal habit for one animal or group may be abnormal for another. Inspection at a distance should be done first. To further examine the animal, it should be restrained so temperature, heart rate and respiratory rate can be take and along with a systematic examination of all major areas of the animal body, noting amount, color, turbidity, odor and consistency of discharges from any of the body openings. Evaluation also includes reviewing external body surfaces, looking for changes in color, size and shape. An important symptom of any animal is pain and it should always be a major concern when evaluating an animal's overall health.

Additional Resources

Taylor, R.E. and T.G. Field. 2010. Scientific farm animal production: Introduction to animal science. 10th ed. Upper Saddle River, NJ: Prentice Hall.

USDA – National Animal Health Surveillance System
https://www.aphis.usda.gov/aphis/ourfocus/animalhealth/monitoring-and-surveillance/sa_nahss/ct_nahss

Assessment

Take assessment online here: http://tinyurl.com/AnSci-AnimalHealth
Download and print the assessment by scanning this QR code or by going to this URL: http://www.tagmydoc.com/Ch38AnSci

39 Responsibility for Animal Health

Major Concept

Animal health is a planned program concerning health maintenance and the prevention of disease in livestock.

Objectives

- Identify the three levels of responsibility for an animal's health
- List the major roles of each level of responsibility in maintaining animal health

Key Terms

- Brucellosis
- Psitticosis
- Rabies
- Tuberculosis

Chapter Resource

 Complementary *full color* illustrations, photos, charts and graphs are available by scanning this QR code or by following this URL: http://www.tagmydoc.com/AS39 This digital resource will enhance your understanding of the chapter concepts.

Animal Health Concerns

- Animal health is a planned program concerning health maintenance and the prevention of disease.

- Three levels of responsibility:

    1. Animal Owner
    2. Veterinarian
    3. Government Officials and Services

    o All are effective when working together.

Animal Owner

- Owners have the immediate responsibility for the animals' health in their possession.

- Owners provide feed, water, shelter, sanitation and health care for the animal.

- In the best position to know when something goes wrong, as the owner/manager should be observing the animal on a regular basis.

- Owners are responsible for obtaining knowledge about their animals from various sources to maintain the level of health desired.

    o Resources include:

        ✓ Veterinarians
        ✓ Cooperative extension services
        ✓ State universities
        ✓ State experiment stations
        ✓ Feed suppliers
        ✓ Other local livestock owners
        ✓ Livestock associations
        ✓ Libraries – books, journals and magazines
        ✓ Internet

Veterinarian

- A professional person whose education is in animal health and is trained in preventative medicine and treatment of health problems.

    o Producers establish an early and continuing relationship with their veterinarian.

    o Provides services and instruction when health problems arise, but may provide these services and instruction to maintain and prevent health-related problems (preventative medicine/practices).

    o Veterinarians have the responsibility of reporting certain diseases to the government, due to public health significance or due to government animal health program regulations.

        ✓ Examples include rabies, anthrax or cholera.

Chapter Resource

Government

- Governmental mandates (regulations/laws) prevent, control and eradicate several specific types of diseases.

    o Animal diseases that can be transmitted to humans include:

        ✓ **Rabies** – a virus transmitted through bites.

- ✓ **Brucellosis** (undulant fever) – bacteria that causes abortion in affected females.
- ✓ **Psitticosis** (parrot fever) – acute or chronic disease characterized by respiratory and systemic infection.
- ✓ **Tuberculosis** – caused by bacteria and usually settles in the lungs.

   o Diseases of economic importance to livestock industries but not transmitted to humans are also subject to regulation.

   o Protection against foreign diseases that could destroy domestic livestock and poultry industries, could become established in this country, is provided through laws and regulations.

- The USDA and the U.S. Customs Service are responsible for the controlling of animal importation and the potential diseases they could bring with them to our country.

Summary

Animal health is no accident. It is a planned program concerning health maintenance and the prevention of disease. The owner is responsible for obtaining knowledge about his/her animals from various sources to maintain the level of health desired. A veterinarian can not only provide services and instruction when health problems arise, but may provide these services and instruction to maintain and prevent health-related problems (preventative medicine/practices). Some animal diseases can be transmitted to man including: rabies, brucellosis, psitticosis and tuberculosis.

Additional Resources

Hayes, J. 1984. Animal health: Yearbook of agriculture. Printing Office, Washington, D.C.

Kahn, C.M. and S. Line. 2010. The Merck veterinary manual. Rathway, NJ: Merck and Co., Inc.

Animal and Plant Health Inspection Service
https://www.aphis.usda.gov/aphis/home/

Animal Health and Veterinary Laboratories Agency, United Kingdom
http://www.defra.gov.uk/ahvla-en/

Animal Welfare Act and Animal Welfare Regulations
https://www.aphis.usda.gov/animal_welfare/downloads/Animal%20Care%20Blue%20Book%20-%202013%20-%20FINAL.pdf

USDA - Animal Health
http://www.usda.gov/wps/portal/usda/usdahome?navid=ANIMAL_HEALTH

Assessment

 Take assessment online here: http://tinyurl.com/AnSci-Responsibility
Download and print the assessment by scanning this QR code or by going to this URL: http://www.tagmydoc.com/Ch39AnSci

40 Disease Causing Agents

Major Concept

Major infectious and non-infectious agents have prescribed methods of control.

Objectives

- Name four general groups of infectious agents causing diseases
- Identify parasites, describe how they may harm the host
- Identify specific ways that infectious agents may gain entrance and do harm to an animal
- List non-infectious agents causing diseases

Key Terms

- Arthropods
- Bacteria
- Disease
- Fungi
- Helminths
- Host-specific
- Lethargic
- Nanometers
- Parasites
- Pathogenic
- Protozoa
- Viruses

Chapter Resource

 Complementary *full color* illustrations, photos, charts and graphs are available by scanning this QR code or by following this URL: http://www.tagmydoc.com/AS40 This digital resource will enhance your understanding of the chapter concepts.

Disease

- **Disease** is any deviation from or interruption of the normal structure or function of any body part, organ, or system that is evident by a characteristic set of symptoms and signs.

- Cause, development and prognosis may be known or unknown.

Infectious Agents Causing Diseases

- Infectious agents causing diseases (**pathogenic**) can be divided into four general groups:

1. Bacteria
2. Viruses
3. Parasites
4. **Fungi** (spore producing microorganisms such as yeast and mold)

- **Bacteria**
  - Most simple form of life.
  - Microscopic one-celled organisms that vary in size and shape.
  - Can be pathogenic (disease-causing) and in other cases are beneficial to man, (common ones are used in the fermentation of vinegar and in cheese making).
  - Generally, live outside the host cell.
  - Sensitive to antibiotic drugs such as penicillin.

Chapter Resource

- **Viruses**
  - An ultramicroscopic (20 to 300 **nanometers** [nm] in diameter), metabolically inert, infectious agents that replicate (makes copies) only within the cells of living hosts, mainly bacteria, plants and animals.
  - Composed of an RNA or DNA core, a protein coat, and in more complex types, a surrounding envelope.
  - Live within the tissue of host cells and therefore cannot be killed with medications that circulate only in body fluids (unlike bacteria).
  - Resistant to antibiotic drugs.
  - Most are highly contagious.

- **Parasites**
  - Organisms living on, in, or at the expense of the host.
  - Parasite infections develop slowly and are difficult to eradicate completely.
  - Animals do not develop immunity, to parasites, although some resistance is apparent between age groups.

- o Some animals do not show many symptoms, others may show symptoms such as dull hair coat, **lethargic** (sluggish) behavior and loss of weight.
- o Most parasites are **host-specific** (only live in certain types of animals).

- Included in this category are **protozoa** (one-celled, mobile organisms with a nucleus), **arthropods** (invertebrate animal with exoskeleton, jointed appendage and segmented body) and **helminths** (worms).

    - o Parasites in this group

        - ✓ Cause tissue damage through migration and consumption.
        - ✓ Absorb nutrients from the host's gastro-intestinal system.
        - ✓ Suck blood or lymph from the hosts body.
        - ✓ Obstruct passages.
        - ✓ Cause nodules or growths internally and externally.
        - ✓ Cause general irritation internally and externally.
        - ✓ Transmit other types of diseases to the host.
        - ✓ Open the body to secondary infection.
        - ✓ Cause serious economic damage to producer because of the general poor health of the stock, related bills and lower sale prices.

Chapter Resource

Life Cycles of Some Common Parasites

- Roundworm (ascarids)

    - o Internal parasites commonly affecting young animals including dogs, cats, horses and hogs.

    - o Cause severe damage due to migration of thousands of larvae through the body tissues during their life cycle.

    - o Eggs may survive for years under proper weather conditions.

    - o Reach maturity in about three months.

    - o Eggs are passed out of the body through feces which hatch on ground or are consumed by another animal.

    - o In larval form can be ingested, after which they penetrate the intestinal wall and burrow into the bloodstream.

- Larva move through blood to liver, in about a week and eventually reach the lungs.
- Larva grow and develop in the lungs, causing irritation and the creation of a severe cough.
- Larva are finally coughed up, re-swallowed, and return to intestines for two months where larva mature and reproduce and pass eggs out of the feces.
- Then the cycle begins again when the larva (from the eggs) are eaten by another animal.

Chapter Resource

- Tapeworms

  - Are internal parasites commonly affecting cattle, sheep, dogs, cats and horses.
  - May have more than one intermediate host such as the flea, grass mite or rabbit.
  - The flea is the most common intermediate host and can live for two years without feeding (and as long as the flea lives, so do the worm larvae).
  - Tapeworms have two parts:
    1. Head (scolex) which attaches to intestines and starts producing proglottids which are passed out with the feces.
    2. Segments (proglottids) which are egg-filled sacs that are passed out and can be seen in fresh feces (they look like bits of rice).

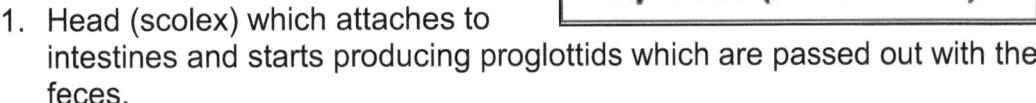

Tapeworm segment as it appears in stool (25x actual size)

Tapeworm (2x actual size)

  - Intermediate hosts or other animals ingest eggs and infective larva by gnawing, eating grass or by consuming another animal (for example, dogs eat rabbits) or their feces.

- Mosquitoes

  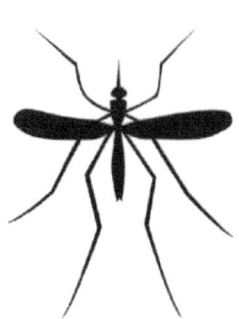

  - An external parasite of all warm-blooded animals.
  - Carry many diseases such as encephalitis, malaria and fowl pox.
  - May also be the intermediate host or vector for other parasites.

- During warm weather can move completely through the life cycle (from egg to adult) in only one week.
- Lay eggs directly on the water or on dry surfaces.
- Eggs hatch to larval form once they are moistened.
- Larva change to pupa form in the water.
- Finally, metamorphosize into adults and the cycles begins again.

- Bot Flies

  - An external parasite of horses.
  - Life cycle, from egg to adult, takes one year.
  - Lay eggs on the horse's leg; (up to 500) after which the horse licks or scratches the area with its mouth, the eggs get into the mouth and the moisture stimulates eggs to hatch.
  - Larva develops in mouth at base of the teeth and gums.
  - Larva is swallowed and migrates to the stomach where they attach themselves to the stomach wall for 8 to 10 months.
  - Larva changes to the pupal stage (non-feeding stage) and pass to ground with feces.
  - Pupa burrow into the ground until adult fly emerges during warm weather in about 4 to 8 weeks and the cycle begins again.

Chapter Resource

Fungi

- Large group of spore-producing organisms that includes microorganisms such as yeasts and molds.

- Some spores cause disease.

- Common, familiar disease is ringworm associated with many farm animals, including cattle, pigs and horses.

  - Can be transmitted from animals to humans.

- Many could be considered parasites on plants and animals.

Non-Infectious Agents that Cause Disease

- Chemicals and poisons

- Poor nutrition

- Injuries

- Physical stress found during breeding, conception, lactation and birth.

- Any other kind of stress or trauma that weaken the animal and its immune system, making it more susceptible to other secondary infections.

Chapter Resource

Summary

Disease is any deviation from or interruption of the normal structure or function of any body part, organ, or system that is evident by a characteristic set of symptoms and signs. Infectious agents causing diseases (pathogenic) can be divided into four general groups: bacteria, viruses and parasites and fungi. Included in the fungi category are protozoa, arthropods and helminths. Common parasites are roundworm, tapeworm, bot flies and mosquitoes and their lifecycles vary. Fungi are a large group of spore-producing organisms that includes microorganisms such as yeasts and molds. Some spores cause diseases that can be transmitted from animals to humans. Non-infectious agents that cause disease include chemicals and poisons, poor nutrition, injuries, physical stress and any other kind of stress or trauma that weaken the animal and its immune system, making it more susceptible to other secondary infections.

Additional Resources

Internal Parasites Common in Horses
http://www.drsfostersmith.com/pic/article.cfm?d=508&articleid=1602

Internal Parasites in Beef and Dairy Cattle
http://www.uaex.edu/Other_Areas/publications/PDF/FSA-3045.pdf

Poultry for the Consumer
http://extension.msstate.edu/agriculture/livestock/poultry

The Complete Guide to Chicken Parasites
http://www.thehappychickencoop.com/guide-to-chicken-parasites/

Assessment

 Take assessment online here: http://tinyurl.com/AnSci-Pathogenic
Download and print the assessment by scanning this QR code or by going to this URL: http://www.tagmydoc.com/Ch40AnSci

41 Development and Types of Immunity

Major Concept

Animals resist disease through two lines of defense which can be enhanced by good management.

Objectives

- Identify the different components of the two lines of defense against disease
- Name three types of immunity
- Identify how antibodies are formed
- List five predisposing conditions that could cause animal health problems

Key Terms

- Acquired immunity
- Active immunity
- Acute
- Chronic
- Cilia
- Detoxifies
- Immune response
- Immunoglobulins
- Lymphatic system
- Natural immunity
- Passive immunity
- Predisposing

Chapter Resource

 Complementary *full color* illustrations, photos, charts and graphs are available by scanning this QR code or by following this URL: http://www.tagmydoc.com/AS41 This digital resource will enhance your understanding of the chapter concepts.

Healthy Body against a World of Disease

- Two lines in the body's defense against disease organisms.

- First line of defense includes:

  o Skin which is covered by hair, feathers, etc., and is thick and able to sweat (some species must pant).

- Mucous membranes which are the sticky, protective coating that trap pathogens from further penetration (e.g. in the nose) and the tearing that washes pathogens out of the eyes.

- Gut which is difficult for pathogens to pass through because of the acids that are produced by digestive processes.

- Cell wall which is often mucous coated and may have **cilia** (a short, microscopic, hair like vibrating structure) to protect the cell.

- Coughing which rids the body of disease organisms which may have been swallowed and the mucus they may be trapped in.

- Flushing effect of urination, which helps clean the ureter (the duct by which urine passes from the kidney to the bladder) by washing pathogens out.

• Second line of defense includes the white blood cells and the lymphatic system, the liver and antibodies.

- **Lymphatic system** which filters pathogens and other undesirables out of the body's lymphatic system.

- Liver which filters blood and **detoxifies** (removes toxic substances or qualities) poisons.

- Antibodies also called **immunoglobulins** are any of various proteins produced in the blood in response to the presence of an antigen (pathogen).

    ✓ By becoming attached to antigens on infectious organisms, antibodies can render them harmless or cause them to be destroyed.

    ✓ Once pathogens have breached the body's line of defense, they generally multiply and attach and destroy body tissues.

Chapter Resource

Immunity

• Condition of being able to effectively combat infection.

• Can generally be obtained in two different ways:

    1. A body's response to a disease in which the body generates antibodies to help protect against reinfection.

    2. Immunization (vaccination) with a compound developed to stimulate an **immune response** (i.e. - generate antibodies to protect against specific diseases).

- Usually, if immunity is gained in this way, there is little or no sickness, which helps maintain rates of gain and production.
- The animal does not need a recovery period during which time it may be off feed.

Immunity or Resistance to Disease – Two Groups

- **Natural immunity** refers to the protection an animal has when it is born.
  - Inherited by an animal from its parents.
  - May be an attribute of the species, breed or individual (horses do not get hog cholera).

- **Acquired immunity** is associated with the presence of antibodies from another immune animal or from exposure to the disease.

Chapter Resource

- Two types of acquired immunity:
  - **Passive immunity** is acquired by transferring of antibodies from an immunized animal to an unimmunized one.
    - ✓ Immunity is immediate and lasts as long as antibodies remain in the body, usually 3 to 6 weeks.
    - ✓ Usually the blood serum (fluid only, no cells) of one animal is collected and used to induce passive immunity in another.
    - ✓ Young or newborn animals receive passive immunity from colostrum obtained from their mothers.
    - ✓ The newborn of many farm species can absorb the antibody proteins directly from their dam's milk into their blood stream (without first digesting those proteins) for about 24 hours after birth. After which time the stomach begins to work and all ingested proteins are broken down, including the antibodies.
  - **Active immunity** is acquired through direct contact with the specific disease-causing organism that causes the body to develop antibodies to combat invasion.
    - ✓ Can be developed after contracting and recovering from a specific disease.
    - ✓ Can also be developed after inoculation of the animal with a mild form (usually a chemically or physically altered) of the disease.
    - ✓ Active immunity is relatively long-lived (often is life-long).

Disease Development

- Disease activity results in changes in the tissues invaded.
    - Typical signs of infection are:
        - ✓ Redness due to increased blood flow to the injured area
        - ✓ Swelling
        - ✓ Localized heat around the injury and/or increased body temperature
        - ✓ Pain resulting from excessive pressure on tissues due to the swelling

- Speed at which a disease attacks an animal is termed one of two ways:
    1. **Acute** – relatively sudden appearance of symptoms (within 24 hours).
    2. **Chronic** – that which develops more slowly, lingers and will frequently reappear.

- **Predisposing** (inclined to) conditions that bring on disease are:
    - Overwork
    - Exposure to cold, heat, rain, snow, sun, humidity, other animals and parasites
    - Long shipments
    - Weaning
    - Injury
    - Management practices such as docking, castration, branding, tagging, dehorning, etc.
    - Access to stagnant water
    - Spoiled feeds and poisonous plants

Chapter Resource

Ways Diseases Spread

- Direct contact
    - Animals rubbing against each other (ringworm), sexual contact (venereal infections), or by mucus (respiratory diseases).
- Contact with non-living objects

- Fence posts, trucks, feeders, needles
- Infection from soil
  - Tetanus, blackleg
- Infection from food or water
  - Influenza (viral infections)
- Airborne infection
  - Coughing and sneezing of crowded animals on each other.
- Infection from blood suckers
  - Mosquitos and flies carrying such diseases as encephalitis.
- Infections from organisms normally in the animal body that only becomes dangerous when animals' defenses are weakened by some other health or stress-related problem/s predisposing them to disease.

Chapter Resource

Good Management for Disease Prevention

- Be alert for signs of disease and conditions which can cause stress and strain.
- Provide clean, disinfected quarters, free from draft.
- Provide adequate ventilation and plenty of sunlight.
- Provide proper drainage of holding areas, barns, free stalls, etc. to help maintain the driest area possible.
- Protect them from the sun, rain and wind without overcrowding.
- Practice rigid sanitation and manure removal procedures.
- Provide a well-balanced diet.
- Get accurate diagnosis of health problems immediately so that treatment can be provided.
- Avoid unnecessary stress and strain.
- Buy disease-free stock from healthy herds and flocks.

- Isolate new animals for a period (to be sure they are healthy) before introducing them to your herd or flock.

- Follow a set vaccination program.

- Be cautious of visitors from other operations as they may carry disease pathogens on shoes and clothing.

- Dispose of dead animals immediately.
    - If cause of death is unknown, contact your veterinarian and have an autopsy performed.

Summary

Animals possess two lines of defense against disease organisms. Once pathogens have breached the body's line of defense, they generally multiply and attach and destroy body tissues. Immunity is the condition of being able to effectively combat infection and is obtained by a body's response to a disease in which the body generates antibodies to help protect against reinfection, or by immunizations. Immunity or resistance to disease is either natural or passive. Active immunity is acquired through direct contact with the specific disease-causing organism that causes the body to develop antibodies to combat invasion. Disease activity results in changes in the tissues invaded. Predisposing conditions that bring on disease include: overwork, exposure to cold, heat, rain, snow, sun, humidity, other animals and parasites, long shipments, weaning, injury, management practices, access to stagnant water and spoiled feeds and poisonous plants. Diseases can spread by: direct contact, contact with non-living objects, infection from soil, infection from food or water, airborne infection, infection from blood suckers and infections from organisms. Good management will do much to help animals resist disease.

Additional Resources

Taylor, R.E. and T.G. Field. 2010. Scientific farm animal production: Introduction to animal science. 10th ed. Upper Saddle River, NJ: Prentice Hall.

Disease Resistance in Cattle
http://extension.usu.edu/files/publications/factsheet/AH_Beef_23.pdf

Immune Defense against Microbial Pathogens
http://www.textbookofbacteriology.net/immune.html

Innate Immunity in Bovine
http://tinyurl.com/ms4kpuh

Assessment

 Take assessment online here: http://tinyurl.com/AnSci-Development
Download and print the assessment by scanning this QR code or by going to this URL: http://www.tagmydoc.com/Ch41AnSci

42 Vaccination and Administration of Biologic Agents

Major Concept

Vaccinating livestock against diseases contributes to a good health management program.

Objectives

- Identify five general types of biologics used in vaccinations
- List three methods of applying treatments
- Name three types of injections

Key Terms

- Antiserum
- Aspirate
- Bacterin
- Biologicals

- Confer immunity
- Modified Live Viruses(MLV)
- Toxoids

Chapter Resource

 Complementary *full color* illustrations, photos, charts and graphs are available by scanning this QR code or by following this URL: http://www.tagmydoc.com/AS42 This digital resource will enhance your understanding of the chapter concepts.

Vaccination

- Injection of some agent (such as **bacterin** (a suspension of killed or attenuated [weakened] bacteria for use as a vaccine), serum or vaccine) into the animal for developing disease resistance (immunity).

- By itself, a vaccine does not **confer immunity** (grant or provide resistance of an infection or toxin by the action of specific antibodies or sensitized white blood cells); it only stimulates an animal's own immune system into action.

- **Biologicals**, biologics or biological products are all terms that cover materials used for vaccination.

- Biologic agents used in developing disease resistance/immunity:

    o Living viruses are used to stimulate antibody production against a specific virus without the animal contracting the disease.

    o Living (unmodified) viruses are actual living viruses that are administered to (usually) young animals when their expected reaction will be mild.

    o A dangerous virus, may be administered with an **antiserum** (to help the animal's body fight the virus until the animal's system makes its own antibodies) to help stimulate the immune response.

    Chapter Resource

    o Living viruses confer an effective and lasting immunity.

    o Killed or inactivated viruses are administered to a susceptible animal to stimulate active immunity (body generates its own immunity) in the animal.

        ✓ Although the response is usually very strong; the process may need to be repeated every 12 months.

    o **Modified Live Viruses** (MLV) are products which contain a live virus but have been changed or modified to not cause the disease but still stimulate antibody formation against the disease.

        ✓ Modified virus is collected from an unnatural host (other than the type of animal that normally is the host for the pathogen).

        ✓ An example is hog cholera – it can grow in an embryonated egg (the unnatural host), collected, processed and used to vaccinate the host.

    o A bacterin is made of the killed preparation of the disease-causing bacteria; it will not cause disease or spread, but stimulates antibody formation.

        ✓ These agents have a relatively lower power to confer immunity and must be re-administered at least once every 12 months.

    o **Toxoids** are an inactivated, altered toxins (the poison that is produced by pathogenic bacteria) used to stimulate immunity.

        ✓ Toxoids provide long-lasting protection against problems such as tetanus and enterotoxemia.

    o Antisera including antitoxins are blood serums containing antibodies which are injected into unprotected animals, stimulating antibody formation.

        ✓ This creates "passive" (it is received by using antibodies from the injection – instead of generating any) immunity which is immediate, but short-lived.

- ✓ An example is tetanus antitoxin.

Administering Biologic Agents

- Read carefully and understand all the information on the medication label.
- Never use medication on animals other than as prescribed in the directions.
- Be aware of brand versus generic names for medicines.
- Be aware of side effects.
- Use the proper amount of medication in relation to the size of the animal.
- Be sure to follow the storage instructions.
- Check the expiration date.
- Be certain that you have read the information and are sure of the best method of administration of the compound.

Methods of Treatment Application

- By sprays and dusts - used often with poultry
- Through inhalation
- By mouth (oral) – pills, tablets, paste, food additives, etc.
- In drinking water
- By absorption – painting the anus, dipping the entire animal, or capsules in the ear
- By injection

Chapter Resource

Methods of Injection

- Intramuscular (IM) injections:

  o Most commonly used for antibiotics, vaccines and wormers.

  o Are easy to administer in large muscles of neck or thigh.

  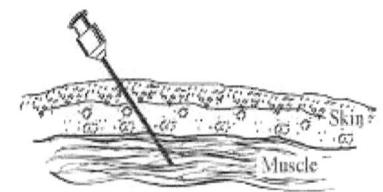
  Intramuscular Injection

  o Must be injected into the muscle tissue, not a blood vessel, hence the need to "**aspirate**" (pull back on the syringe plunger to be sure the needle isn't in a blood vessel) before injecting the compound.

- Subcutaneous (Sub-Q) injections are:

  o Injection of the compound (fluid) directly beneath the skin, not in the flesh or a blood vessel.

  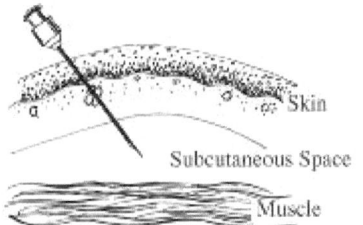
  Subcutaneous Injection

  o Easy to do where the skin lies in loose folds such as under the mane in horses, on the back of the neck in dogs and behind the ear in swine.

- Intradermal injections (ID):

  o Made between the skin layers, not beneath it.

  o Done with a very fine gauge needle.

  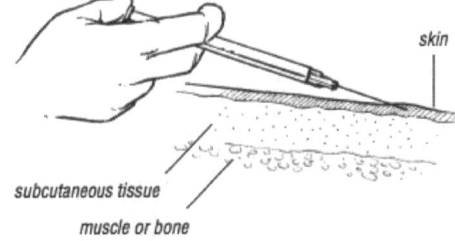

- Intravenous injections (IV):

  o Injected directly into a blood vessel (using ones that are easily identified and accessible)

  o Action of drug is usually almost immediate (15 to 60 seconds) after injection. Therefore, solutions should be injected slowly to avoid side effects or the "shock" of too much at once.

Summary

A vaccination is the injection of some agent (such as bacterin, serum or vaccine) into the animal for the purpose of developing disease resistance (immunity). Biologicals, biologics or biological products are all terms that cover materials used for vaccination. Living viruses are used to stimulate antibody production against a specific virus without the animal contracting the disease. Living viruses confer (grant or provide) an effective and lasting immunity. Modified live viruses (MLV) are a product which contains a live virus but has been changed or modified to not cause the disease but still stimulate antibody formation against the disease. A bacterin is made of the killed preparation of the disease-causing bacteria; it will not cause disease or spread, but stimulates antibody formation. Use the proper amount of medication in relation to the size of the animal. The methods of injection include: intramuscular, subcutaneous, intradermal and intravenous.

Additional Resources

Hayes, J. 1984. Animal health: Yearbook of agriculture. Printing Office. Washington, D.C.

Kahn, C.M. and S. Line. 2010. The Merck veterinary manual. Rathway, NJ: Merck and Co., Inc.

Developing a Livestock Vaccination Program
https://cals.arizona.edu/backyards/sites/cals.arizona.edu.backyards/files/p6-8.pdf

Developing Trends in Livestock and Poultry Vaccines
http://www.isca.in/AVFS/Archive/v1/i3/5.ISCA-RJAVFS-2013-015.pdf

Standard Operating Procedures – Cattle Vaccination
http://tinyurl.com/nzhdmz3

Assessment

Take assessment online here: http://tinyurl.com/AnSci-Vaccination
Download and print the assessment by scanning this QR code or by going to this URL: http://www.tagmydoc.com/Ch42AnSci

43 Beef Cattle Industry

Major Concept

Recognize the types of beef cattle enterprises, management and feeding of cattle and calf management.

Objectives

- Name the three types of beef cattle enterprises
- Identify the advantages and disadvantages of spring and fall breeding seasons
- List the most common forages for cattle
- Outline a sample calf management plan
- Outline a pasture management plan
- Identify general guidelines for winter feeding cattle

Key Terms

- Artificial Insemination
- Bloat
- Brood
- Commercial herds
- Creep feeding
- Drylots
- Estrus
- Forages
- Heifers
- Herds
- Hundred-Weight
- Insemination
- Lactating
- Prostaglandin
- Ration
- Seedstock herds
- Steroid

Chapter Resource

 Complementary *full color* illustrations, photos, charts and graphs are available by scanning this QR code or by following this URL: http://www.tagmydoc.com/AS43 This digital resource will enhance your understanding of the chapter concepts.

International

- United States ranks 4th in total number (99 million) and first in carcass beef produced (20 billion lbs).

- Other important beef producing countries include: Brazil, China, former Soviet Union, Argentina, Mexico, Colombia, Australia and France.

In the United States

- Cattle inventory: 93.5 million, up 1.8% from January 2016

    - ✓ 31.2 million beef cows
    - ✓ 6.4 million beef replacement heifers in 2017
    - ✓ 9.35 million milk cows

    - o 35 million head calf crop (2016)
    - o 2016 Forecasted Economic impact: $67.56 billion in farm cash receipts for cattle and calves
    - o 43.4 billion lbs of beef harvested under USDA inspection.
    - o Average live weight 1,277 lbs.
    - o Total cash receipts: $68 billion.
    - o Beef cattle are produced in all states in the United States.

- Top 5 states that raise cattle and calves as of Jan. 1, 2017:

    1. Texas
    2. Nebraska
    3. Kansas
    4. California
    5. Oklahoma

- Check out this website for updated statistics:
  http://www.beefusa.org/beefindustrystatistics.aspx

Chapter Resource

Type of Beef Cattle Enterprises

- Three types of enterprises

    1. Cow-calf
    2. Stocker
    3. Feedlot

- Cow-calf program

    - o Maintains a cow herd and produces calves, usually sold at weaning as feeder cattle.
    - o Based on pasture which is available during the spring, summer and fall.

- Principal **forages** (plant material, leaves and stems) are fescue, bermudagrass, bluegrass (mountains), orchardgrass and clovers.

- Some harvested forage or by-products (corn silage, hay, straw, crop residues) are required in the winter.

- Enterprise requires much land for feed production.

- Grains and supplements used only to the extent they are required to balance the **ration** (a fixed amount of a commodity).

- Feeder cattle

  - Less risk in this program and it is the safest for new producers to follow.

  - Calves are usually born during the winter and are sold directly off the cow the following fall (September, October) at about 6 to 9 months of age and weighing about 450-600 lbs.

Chapter Resource

- Stocker cattle

  - Calves or older animals maintained, often on pasture or rangeland, to increased weight and maturity before being placed in a feedlot.

  - Cattle fed for various periods on such feeds as pasture, corn silage, etc.

  - Fed for growth not finish (average expected rate of gain, 1.0-1.25 lbs/day).

  - Sold at some later date to a cattle feeder that places them in a feedlot for fattening.

- Feedlot

  - Producer purchases stocker cattle and feeds them to market weight.

  - Cattle finishing requires more concentrate and less roughage than the other enterprises.

  - Usually requires less land and is a rather speculative enterprise.

- Most important methods of finishing are:
  - ✓ Drylot or high intensity production (much grain, little roughage, little land with large investment).
  - ✓ Grain-on-grass. Uses pasture in a finishing program (less concentrate, more pasture).
- Reasons most cattle move to Mid-western **drylots** (no pasture; daily feed and water is provided by the caretaker) are:
  - ✓ Southeast grain deficient area
  - ✓ Urbanization
  - ✓ Poor pastures
  - ✓ Lack of marketing opportunities

Chapter Resource

- Most feedlots are in the Mid and Southwestern US where the grain is produced.

Beef Cattle Breeds

- Oklahoma State University's Breeds of Livestock website
  - http://www.ansi.okstate.edu/breeds/
- Background and distinguishing characteristics and pictures of the most numerous beef breeds in the United States are covered in the above website.
- Common breeds include:
  - Angus
  - Hereford
  - Limousin
  - Charolais
  - Simmental
  - Red Angus
  - Beefmaster
  - Brangus
  - Brahman
  - Santa Gertrudis
  - Gelbvieh
  - Salers

Cow-Calf Management

- Breeding and calving season
  - Heifers (young cows before calving) and cows can be bred to calve during the spring or fall.
  - In the spring calving systems, cows calve in February and March and heifers calve in January and February.

- o In fall calving systems, cows calve in November and December and heifers in October and November.
- o Availability of markets, feed supply, labor, temperature and rainfall should be considered when deciding which breeding season to be used.

- Spring calving season: Advantages

    - o Calves old enough to use the cow's abundant milk supply when pastures lush.

    - o Cow fertility usually higher during spring and early summer.

    Chapter Resource

    - o Little or no supplemental or additional feeding required if marketed at weaning.

    - o Fewer problems with calf diseases.

    - o Spring calves usually have higher 205-day adjusted weaning weights.

    - o Weather conditions generally more favorable for healthy calves.

- Spring calving season: Disadvantages

    - o Labor needed when it is least available.

- Fall calving season: Advantages

    - o Calves sold the next fall are heavier but must be fed from weaning to market.

    - o Labor more readily available for late fall calving.

- Fall calving season: Disadvantages

    - o Must provide supplemental feed for cow in drylot for high milk production.

    - o Although calves are heavier the next fall, price per **hundred-weight** (a unit of mass defined in terms of the pound) is often lower.

Breeding

- April and May for spring calving.
- January and February for fall calving.

- Heifers are bred beginning 30 days before the older cows.

    o Gives the producer more time with them at calving and gives them a better chance to rebreed as 2-year old's suckling their first calf.

- Breeding season management

    o Most beef cows are pasture bred, where bulls run with the cows during the breeding season.

    o Yearling bulls should not be pasture bred to more than 10 females in a 90-day breeding season.

    ✓ 2-year old bulls 20-25 cows.
    ✓ 3-year olds and older bulls 30-40 cows.

- A high level of reproductive performance required and a goal of 90% calf crop should be set.

Artificial Insemination (AI)

- The process of collecting semen from a bull and manually placing or depositing into the cervix of a cow.

- Requires high levels of management.

- Lower calf crop percent will result if attention is not given to:

    o Heat checking

    o Semen quality and handling

    o Insemination technique; for example: is the technician properly trained?

- Synchronization of **estrus** (a recurring period of sexual receptivity and fertility in many female mammals; heat) with a **prostaglandin** (cyclic fatty acid compounds with varying hormone like effects, notably the promotion of uterine contractions.) or **steroids** (any of a large class of organic compounds with a characteristic molecular structure containing four rings of carbon atoms) only for those cattle producers with excellent management skills. Should be used in conjunction with AI.

Crossbreeding

- Recommended for all commercial producers.
- Crossbred **brood** (mother) cows should be used whenever possible.
  - Crossbreeding system used should be simple and adaptable to the herd size.
  - Crossbreeding improves reproductive performance, calf vigor and survival.

Replacement Heifers

- Selected from performance tested females.
  - Heavy 205-day adjusted weights
  - Heavy yearling weights
  - Large frame size
  - Sound feet and legs
  - Quiet disposition

Chapter Resource

- Put on a good growing ration after weaning.
- Overfeeding increases costs and deposits undesirable fat.
  - May reduce future milk production.
- Bred to drop their first calf at about 2-years of age.
  - Gestation length approximately 285 days; heifers should be bred at about 14-15 months of age.
- Yearling heifers are not bred to large size breeds noted for sire-influenced calving difficulty.

Brood Cow Characteristics

- Strong maternal traits such as mothering ability.
- Adequate milk production
- Excellent fertility

- Sound feet and legs
- Adaptable to local environment and management.

Bulls

- Performance tested and acclimated to farm conditions.
- A breeding soundness exam.
    - Semen quality
    - Before start of the breeding season

Calving Season Management

- Calf beef cows in a maternity lot (clean pasture).
    - Not too large and is free of natural hazards such as ponds, creeks and ditches.
    - Windbreaks such as trees or hedges usually provide all the shelter that is needed.
    - If not on pasture and must be inside, clean bedding is a must for the prevention of scours.
- Check cows frequently and be on hand when the calf is born.
- Give assistance only when needed.

Chapter Resource

- See that the calf nurses within 1-2 hours after birth.
- Treat the cord with iodine and identify by ear notch, ear tag or tattoo.
- After calving, move the cow and calf from the maternity lot to a new clean pasture.

Calf Management

- Castrate all bull calves in **commercial herds** (large group of animals kept together as livestock) and all undesirable purebred bull calves in **seedstock herds** (breeding cattle typically registered with a breed association).
    - Recommended age, by 2 months, earlier if practical.

- Methods

  - Knife – Best method
  - Burdizzo (clamp) – Not recommended
  - Elastrator (rubber band) – Do it when young

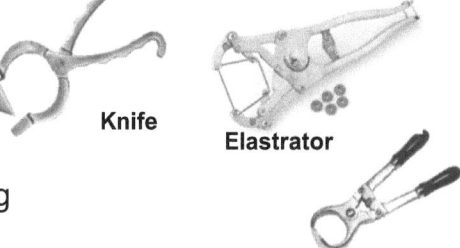

- Castration removes the source of testosterone (and the ability to reproduce).

  - Responsible for secondary sex characteristics and libido.

- Steers are easier to handle and develop more desirable carcasses.

- If bull calves are not castrated, heifer calves and cows may be bred too early or out of season.

- Feeder calf sales will not accept bull calves.

- Dehorn all commercial cattle.

  - Calves should be dehorned by 2 months (usually at the same time they are castrated).

- Dehorning methods, and approximate age when they can be used:

  - Genetic by using a polled bull (very desirable method)
  - Caustic potash (birth to 2 weeks)
  - Electric (hot iron) dehorner (birth to 3 months)
  - Dehorning spoon or tube (birth to 3 months)
  - Barnes dehorner (2 weeks to 6 months)
  - Clippers, saw – for older cattle

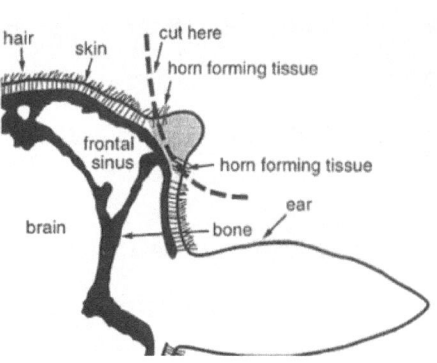

- Horned cattle

  - Require more feeder and barn space.
  - Fight more, have more injuries.
  - Sell at lower price due to bruises.

- o More dangerous to handle.

- Worm calves as needed.

  - o May not be required if good management and sanitation practices are followed.

- Calf vaccination schedule for your area.

  - o Such diseases are blackleg and malignant edemas are serious problems in some places.

- Plenty of shade and fresh water during the summer.

- For injury, castration or dehorning during hot weather; use a fly repellent.

- **Creep feeding** (supplemental feeding) calves with concentrates (grains) pays if:

  - o Pastures become short

  - o Cows are poor milkers

  - o Calves are to go directly to the feed lot at weaning.

    - ✓ Creep feeding does not usually pay if cows are good milkers and pastures are of good quality.

- Weaning – allow cow at least a 60-day dry period before next calving.

  - o Most calves are weaned in the fall at 6-9 months of age.

Chapter Resource

Summer Management: Cow-Calf Enterprise

- Pasture the most important items in the feed program.

- Cows give more milk and wean heavier calves if the pasture contains a legume.

- Legumes are palatable and increase milk production and calf gains.

  - o High in protein and calcium

- When adequate pasture, concentrates are not needed for **lactating** (milking) beef cows.

Common Forages for Beef Cows

- Cool-season perennials – most growth occurs in spring and fall.

    - Tall fescue
    - Bluegrass
    - Orchard grass
    - Perennial ryegrass
    - Clover (legume)
    - Alfalfa (legume)
    - Birdsfoot trefoil (legume)
    - Lespedeza (legume)

- Warm-season perennials – most growth occurs in summer.

    - Bermudagrass
    - Switchgrass
    - Flaccidgrass

- Winter annuals – must be planted each year. Most growth occurs in early spring and late fall.

    - Annual Ryegrass
    - Cereal grains – rye, oats, wheat, barley
    - Arrowleaf clover (legume)
    - Crimson clover (legume)

    Wheat

- Summer annuals – must be planted each year. Most growth occurs in late spring and summer.

    - Pearl millet
    - Tifleat millet
    - Sorghums – forage sorghum, sudangrass, sorghum – sudan hybrids.

Pasture Management

- Pastures fertilized according to soil test and local conditions.

- Pastures clipped at intervals for weed control and for more uniform grazing.

- Provide plenty of shade and fresh water for beef cattle on pasture.

- Fence off ponds and marshy areas to control diseases such as leptospirosis and internal parasites.

- Observe cattle frequently for **bloat** (abnormal swelling of gas in the rumen) if grazing legumes at the start of the grazing season.

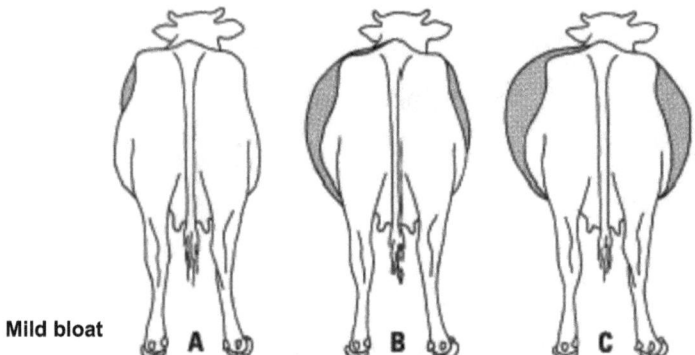

Mild bloat A   B   C

- Isolate newly purchased animals for at least 3 weeks.

- Isolate and treat sick and injured animals.

- Use chemically treated ear tags for fly control.

- Treat for internal and external parasites as needed.

- Do not over-graze pastures during dry weather.

- Use emergency rations (greenchop, hay silage, summer grazing crops, etc.).

- Palapate for pregnancy and cull no-breeders when calf is weaned.

- Feed salt and minerals free choice on pasture.

Chapter Resource

Rules of Thumb for Winter Feeding Breeding and Stocker Cattle

- Dry cows – 1.7 - 2.0 lb hay or equivalent per 100 lbs body wt. which supplies TDN.

    o If hay contains 30% or more legume, or is high quality grass, no protein supplement is needed.

        ✓ Hays are approximately 90% dry matter.

    o No protein supplement is needed.

    o Corn silage is high in moisture (60-70%) and more pounds must be fed.

- Lactating cows – 2-2.5 lbs hay or equivalent per 100 lbs body wt which supplies TDN.

- - Growing cattle (replacement heifers, bull calves, stocker cattle)
    - ✓ Require about 2.5 lb of hay or equivalent per 100 lbs body wt to gain .75 to 1.0 lb per day.
    - ✓ With hay rations, some grain (energy) supplementation may be required if higher gains are necessary.

- Corn silage low in calcium.

- 0.1 lb ground limestone per head per day should be fed (spread over silage).

- Limestone (feed grade) and urea (a compound high in nitrogen which can be converted to NH_3) can be added to good corn silage (high grain content) at ensiling.

- Limestone supplies calcium (Ca) and the urea supplies nitrogen (N) which the microorganisms in the rumen used to make protein.

- With all rations

  - Loose salt should be fed free choice.

  - Mineral mixture

Chapter Resource

- Two parts steamed bone meal (or dicalcium phosphate).

- One part trace mineral salt should also be fed free choice.

Summary

In the world, the United States ranks 4th in numbers of cattle (99 million) and first in carcass beef produced (20 billion lbs). In the United States, 89 million head of cattle are worth over $2 billion. There are three types of beef cattle enterprises – cow-calf, stocker and feedlot. There are many breeds of cattle – the most common breeds are listed on the Oklahoma State University website. Cow-calf management involves breeding and calving seasons; artificial insemination; cross-breeding programs; replacing heifers; characteristics of bulls and brood cows; and managing calving season. Good calf management is important for a healthy calf crop along with good pasture management for cows, heifers and bulls.

Additional Resources

Taylor, R.E. and T.G. Field. 2011 Scientific farm animal production. 10th ed. Upper Saddle River, NJ: Prentice Hall.

Beef Industry Statistics
http://www.beefusa.org/beefindustrystatistics.aspx

Common Cattle Breeds
http://www.livingthecountrylife.com/animals/livestock/16-common-cattle-breeds/

Environmental Protection Agency – Beef Production
http://www.epa.gov/agriculture/ag101/printbeef.html

Penn State Extension – Basic Beef Production Guidelines
http://extension.psu.edu/animals/beef/production/articles/basic-beef-production-guidelines

Assessment

Take assessment online here: http://tinyurl.com/AnSci-BeefCattle
Download and print the assessment by scanning this QR code or by going to this URL: http://www.tagmydoc.com/Ch43AnSci

44 Dairy Cattle Industry

Major Concept

The significant dairy cattle industry and its products in the U.S depend on cattle selection and breeds, breeding techniques and reproductive management.

Objectives

- Name four breeds of dairy cows and identify the top breed
- List three trends in the dairy industry
- Identify three important issues/considerations in dairy cow management
- List three important features of reproductive management
- Identify DHI and NCDHIP

Key Terms

- Artificial Insemination
- BST
- Freshened
- Gomer bulls
- Heritability
- Lactation

Chapter Resource

Complementary *full color* illustrations, photos, charts and graphs are available by scanning this QR code or by following this URL: http://www.tagmydoc.com/AS44 This digital resource will enhance your understanding of the chapter concepts.

Place of Dairy Cattle in U.S. Agriculture

- For the most current information on the U.S. Dairy cattle industry:

  o http://www.ers.usda.gov/topics/animal-products/dairy/background.aspx#.UtR1fvRDuSo

- Dairy products provide over 11% of all yearly cash receipts from agriculture.

  o Second in animal industries behind beef and just slightly ahead of poultry and eggs.

- Dairy products generally account for 20-25% of animal agriculture's share.

- U.S. dairy cattle account for less than 1% of the world's cows.
- Dairy cattle produce about 20% of all beef consumed in the U.S. each year.

Purpose of Dairy Industry in the U.S.

- To make use of resources; for example:
  - Two thirds of the feed energy from forages.
  - By-products from agricultural processes also fed; for example: beet pulp, peels, brewer's grains, corn gluten, distillers grains, etc.
  - Dairy cows are very efficient.
  - Milk production is one of most efficient methods to create food.

- Dairy cows provide a way to add value to feeds.

Historical Perspective

- Began evolving after Civil War.
- During the early 1950's operations became larger.
- All time high of dairy cows was 27.8 million in 1945.
- Shrinking in numbers of producers and cows.
- Technological innovation driving modern dairies.

Chapter Resource

Structure and Geography

- Among the least concentrated because milk is perishable.
- Restructuring since about 1950.
  - Trends
  - Regional shift
- More specialized

- o   Large-scale dairies of several thousand cows.
- o   Small herds are decreasing.
- o   Production shifted west to take advantage of warmer climates.
- Three major dairying areas
  - o   Lakes States
  - o   Northeast
  - o   Pacific northwest

Chapter Resource

Trends in Dairies

- California has become the largest milk-producing state; Wisconsin is second.
- In 1991, the number of dairy cows declined to less than 10 million for the first time.
- During the decline, the average amount of milk produced per cow increased, compensating for the decrease in number of cows.
- Most U.S. dairies fall into two major categories:
  1. Family dairy herds
  2. Large commercial dairies

Herd Improvement

- National Dairy Herd Improvement Association (NDHI): http://www.dhia.org/, and the National Cooperative Dairy Herd Improvement Program (NCDHIP), http://naldc.nal.usda.gov/download/CAT87208338/PDF
- Producers receive information from DHI to help determine the best management tool available to the dairy producer.
- Records also used by the United States Department of Agriculture (USDA) for sire evaluation purposes.
- USDA sire summaries (National Dairy Herd Sire Summaries).
- Accomplished by comparing daughters.

- For any producer using **artificial insemination (AI)** (the process of collecting semen from a bull and manually placing or depositing into the cervix of a cow).

Dairy Selection and Breeding Program

- Good cows make more profit.

- Feed for maintenance.

- Goal for the dairy herd breeding program is to produce replacement heifers that will:

    o Produce large volumes of milk.

    o Calve every 13 months.

    o Stay healthy and avoid or minimize disease problems.

    o Have excellent mobility.

    o Have excellent dispositions.

    o Milk out quickly, cleanly and equally.

    o Have excellent herd longevity.

- Performance

    o Sum of inheritance (genetic potential) and environment (management, feeding, climate and so on).

    o **Heritability** (character traits passed from generation to generation) estimates for dairy traits.

        ✓ Range from about 40% to below 20%, with some in between.

    o Dairy selection especially difficult.

- Selection of Sires and Artificial Insemination (AI) Program

    o Knowledgeable selection required.

        ✓ Good AI programs account for genetic progress in the U.S.

    o Producers follow guidelines for selecting sires.

- o Genetic improvement long term proposition.

Dairy Breeds

- Specialized dairy breeds imported into the U.S. and breed associations established between 1868 and 1880.

- Six dairy breeds used in the U.S. include:

  1. Holstein
  2. Brown Swiss
  3. Ayrshire
  4. Guernsey
  5. Jersey
  6. Shorthorn

- Most popular breed

  o Almost 95% of the dairy cows in the U.S. are grade or purebred Holstein.

- Very small percentage of dairy cattle are registered.

- For information on cattle breeds visit the Oklahoma State University breeds of livestock page at http://www.ansi.okstate.edu/breeds/.

Reproductive Management

- Sterility and delayed breeding problems in the industry.

- Cow must have a calf since milk production declines over time.

- Also need daughters that will become replacements.

Chapter Resource

- Realistic goal of breeding the dairy cow: cow must be pregnant within 115 days of calving or the 13-month calving interval cannot be maintained.

- Reproductive efficiency largely tied to nutrition of the dairy cow.

  o Cows will lose weight since they cannot eat enough to increase production during early **lactation** (the secretion of milk by the mammary glands) and maintain body weight to maintain a lactation curve.

  o Demands of lactation can create a problem making good nutrition necessary.

Artificial Insemination

- Most dairy cattle in U.S. are artificially inseminated.

- Heat/estrus detection important

    o Some use of **gomer bulls** (an intact male that has undergone a penile deviation, penile removal, or vasectomy to render him incapable of physically breeding cows) or hormone treated cows.

    o Others use heat detection methods such as chin-ball markers, marking crayons, or pressure-sensitive pads.

        ✓ (http://www.uaex.edu/Other_Areas/publications/PDF/FSA-4004.pdf)

- Learning the physical and behavioral signs of a cow in heat is necessary for dairy producers and managers.

    o A cow will only display the signs of heat for 12 to 18 hours on a 19 to 21-day cycle.

- Dairy cows are bred and **freshened** (calve) year-round.

Nutrition in Dairy Cattle

- Feeding is the single most important factor in profitability.

- Dairy cattle are ruminants.

- Provided different specialized rations.

- Feeding strategy uses grain and other supplemental feeds.

- Large dairies employ the expertise of a nutritionist.

Chapter Resource

Dairy Herd Health

- Herd health programs are a valuable management tool.

- Dairy animals also require treatment and medicines.

    o Often common to find veterinarians providing both.

- General principals of herd health include:

- o Disease prevention
- o Planned health-related procedures and examinations.
- o Sound record keeping and record use.

Bovine Somatotropin (BST)

- BST – A hormone (growth hormone) produced naturally from pituitary gland.
- BST is a dairy issue with broad implications and controversial issues.
- Found in all species.
- Injections of BST found to increase milk production.
- Made using recombinant DNA technology.

Chapter Resource

- First recombinant DNA product to be approved for use in animals in the U.S.
- Long process to gain approval.
- Commercial use of BST approved for use in 1994.
- Commercial name – Posilac®.

Trends and Influential Factors for the Future of Dairying

- Restructuring – trends
  - o Production shifting to larger operations
  - o Cost of new technologies
  - o Follows overall trend in agriculture
  - o Production continues shifting to the Southern Plains, Mountain and Pacific regions

- Trends – Technology
  - o Historically good users of technology
  - o Continuation expected

- Trends – Genetic markers
  - Genetic markers will allow dairy producers to select with greater accuracy the genetics of the next generation.

Food Safety in the Dairy Industry

- Enviable record of safety

- Consumers react to food safety concerns

- Many concerns are without basis

- Chemical residues considered the most important

- Begins with individual diary producers

- Good, defendable, quality practices needed

Chapter Resource

Environmental Concerns for Dairies

- Waste disposal represents an ongoing concern to large dairies.

- Methods of waste disposal include:

  - Lagoons

  - Composting

  - Separating solids

  - Biogas generation

  - Spreading as fertilizer

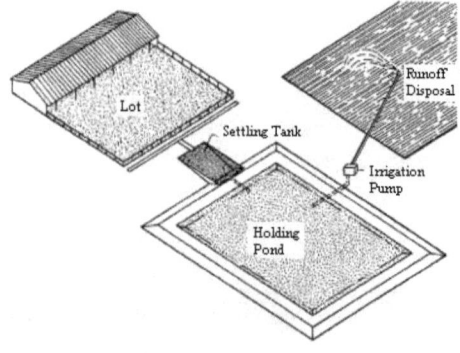

Detention pond for storage of dairy wastewater and livestock-lot runoff

Summary

The dairy industry provides a way to add value to feeds. Dairy products provide over 11% of all yearly cash receipts from agriculture. Highest number of dairy cattle was in 1950 with 27 million. Today it is down to 10 million, but the amount of milk produced from each cow has increased, compensating for the decrease in dairy cattle numbers. The National Dairy Herd Improvement Association (NDHI) and the National Cooperative Diary Herd Improvement Program (NCDHIP) help improve the dairy industry. Areas of importance for good dairy management are: selection and breeding through artificial insemination; performance; breeds, breeding management; nutrition; waste management; and herd health. Dairy operations are become larger and technology has played a major role in their continued success and good food safety records.

Additional Resources

Taylor, R.E. and T.G. Field. 2010. Scientific farm animal production: Introduction to animal science. 10th ed. Upper Saddle River, NJ: Prentice Hall.

Dairy
https://www.ams.usda.gov/market-news/dairy

Dairy Production Systems
http://www.epa.gov/agriculture/ag101/dairysystems.html

North American Dairy Breeds
http://www.ansi.okstate.edu/breeds/cattle/

Assessment

Take assessment online here: http://tinyurl.com/AnSci-DairyCattle
Download and print the assessment by scanning this QR code or by going to this URL: http://www.tagmydoc.com/Ch44AnSci

45 Swine Production Industry

Major Concept

The pork industry and production are an important component of American agriculture and an important part of the American diet.

Objectives

- List three common breeds of pigs in the U.S.
- Name the three types of husbandry systems or farrowing systems
- Outline proper environmental management for a swine operation
- Identify the proper facilities and equipment for a swine operation
- Outline swine feeding and herd health management

Key Terms

- Biosecurity
- Euthanized
- Hand mating
- Husbandry
- Mortality
- Pen mating
- PQA-Plus
- Stress
- Tail docking
- Thermal

Chapter Resource

 Complementary *full color* illustrations, photos, charts and graphs are available by scanning this QR code or by following this URL: http://www.tagmydoc.com/AS45 This digital resource will enhance your understanding of the chapter concepts.

Common Breeds of Swine in the U.S.

- Berkshire
- Chester White
- Duroc
- Hampshire
- Landrace
- Poland China
- Spot
- Yorkshire

Husbandry Systems

- The management and care of farm animals.

- Environmentally controlled buildings

JTF Introduction to Animal Science 270

- Open front buildings
- Outside lots or pastures

Hampshire Pig

Farrowing Systems

- Sow Management
  - Pens
  - Stalls
  - Huts

Chapter Resource

- Systems designed to prevent crushing of piglets.
- Environmental requirements of piglets and sow are different.

Litter Management

- Maintain body temperature (prevent cold stress) and protect from injury.
- Disinfect navels
- Clip needle teeth
- Provide supplemental iron
- **Tail docking** (the intentional removal of part of an animal's tail)
- Castration
- Identification
  - Ear notching
  - Tattoo
  - Ear tag

Ear notching

Nursery Systems

- Weaned at 2 to 4 weeks

- Temperature at least 80°F

- Clean and sanitary

- Piglets fed complete balanced diet

Chapter Resource

Growing and Finishing Systems

- From 50 lbs to 220 to 260 lbs

- Pens hold 15 to 40 pigs

    o Changes with size

- Flooring total or partially slatted

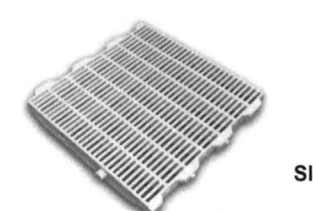
Slatted Flooring

Mating Systems

- **Pen mating** - a cohort of females is brought into the boar's pen and he services them all while they are in the pen.

- **Hand mating** - the boar and one sow (or gilt) at a time are brought together for servicing.

- **Artificial Insemination** - semen is collected from a boar; extended and used by a technician to inseminate many sows.

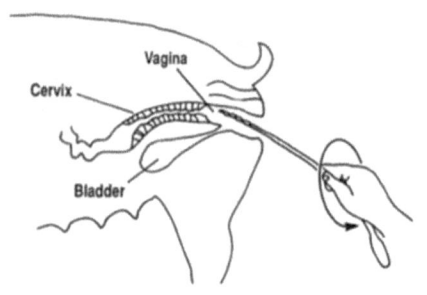

Sow Management during Gestation

- Housing types

    o Gestation stalls

    o Pens for limited numbers

    o Uniform size of females in group

    o Proper amount of floor space for each sow

Good Management Practices

- Observation of pigs and piglets
  - Frequent
  - Know a sick pig

- Plan for emergency, weekend and holiday care.

- Provide proper care at all times

- Keep important names and phone numbers in a convenient location.
  - Veterinarian
  - Feed suppliers
  - Nutrition specialists

- Components of identification and records

- Breeding information

- Feed use

- Growth

- Genetic information

- Farrowing

- Pig **mortality** (the state of being subject to death)

- Marketing

SUGGESTED TARGETS IN THE FARROWING HOUSE	
Total litter size	11.8
Stillbirths	< 5%
Born alive	11.2
Piglet mortality	7%
Pigs weaned	10.4

(Fig.8-4)

CAUSES OF PIGLET MORTALITY AND ACHIEVABLE LEVELS		
Cause of death	The old technology %	The new technology %
Stillborn	8	4
Crushed	7	< 3
Poor viability	1.5	1
Starvation	2	1
Scour	3	0.5
Defects	0.5	0.5
Miscellaneous	2.5	1.5
Total mortality	up to 24.5	11.5
Born live / litter	10.5	11.2
Weaned / litter	8.8	10.4

(Fig.8-5)

PIGLET MORTALITY. FIELD DATA *		
Causes of Piglet Deaths as a % of Total Deaths	% of Pigs Born Alive that Die	
	Actual	Target
Laid on	38	4.1	< 3
Poor viability	18	1.9	1
Miscellaneous	15	1.6	1.5

Handling

- Understand the behavior of pigs

- Reduce undesirable **stress** (strain or tension created by environmental conditions)
  - Proper handling leads to increased growth and reproduction

- Need to know that area of vision for a pig is 300 degrees

- Funnel-shaped pens are <u>not</u> acceptable

- Use abrupt entrance

- Use solid portable panels to control

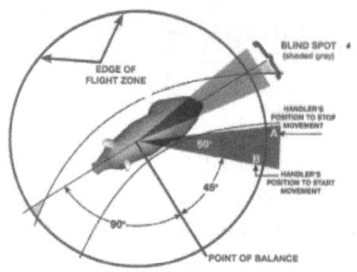

A Pig's Flight Zone, Point of Balance and Blind Spot

Transportation

- Stressful for even healthy animals.

    o Can lose up to 5% body weight

    o Avoid injuries and bruises

    o Problems aggravated by weather

Chapter Resources

- Knowing how and when to ship is an important management practice.

    o Handling sick and injured pigs requires special attention.

- Some animals may need to be **euthanized** (put to death humanely) on farm.

Environmental Management

- **Thermal** (temperature) Control

    o Thermal control is the combined effects of:

        ✓ Air temperature
        ✓ Air speed
        ✓ Humidity
        ✓ Surrounding surface temperatures
        ✓ Insulating effects
        ✓ Physiological state of pigs

- Ventilation Systems

    o Remove: Heat, water vapor, pathogens and air pollutants

    o Influenced mostly by pig numbers

- Air quality issues in swine operations:
  - Dangerous gases (ammonia, hydrogen sulfide, carbon monoxide and dioxide)
  - Dust
  - Microorganisms
- Noise
  - Squealing pigs (100 to 110 decibels)
- Lighting
  - More for producer than pig
  - Pigs less sensitive
- Manure management and sanitation
  - Ensure health
  - Properly manage soil, water and air resources.
  - Minimize dust, pests and parasites.
  - Comply with laws and regulations.

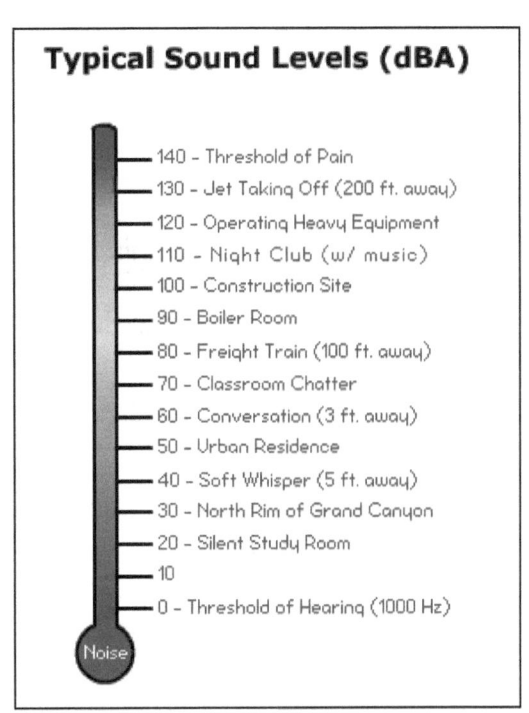

Facilities and Equipment

- Construction guidelines
- Materials:
  - Animal behavior that may lead to structural damage
  - Absence of "stray voltage"
  - Control of pests
  - Manure handling
  - Ease of cleaning and sanitizing surfaces

Chapter Resources

- Lighting
- Moisture and fire resistance

Chapter Resource

- Protection and safety of personnel
- Contact time with wet and corrosive pig wastes and cleaning solutions
- Maintenance of facilities
- Considerations for functional areas of facilities
    - Handling, sorting, weighing, loading and unloading pigs
    - Veterinary examinations, treatment and supplies storage
    - Quarantine of pigs
    - Fenced, penned or enclosed areas with waterers and feeders
    - Water supply for drinking, sanitation, fires and emergencies

    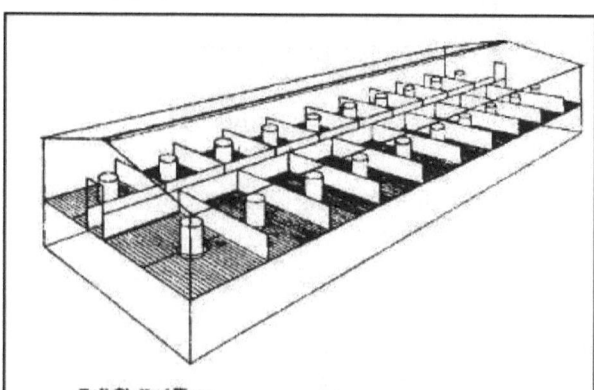
    Full-Slotted Floor

    - Electrical service including an emergency generator
    - Feed storage between deliveries
    - Storage for excreta and contaminated drainage water
    - Storage of equipment used to handle feed and waste
    - Storage of small tools for maintaining and repairing equipment
- Corridors and doors
    - Wide enough for movement
- Floors
    - Non-slip
    - Easy to maintain and clean
- Walls and ceilings

- - Easy to maintain and clean
- Feeders and waterers
  - Size and habit of pigs should be considered
  - Should prevent injuries
- Environmental modification systems
  - Cold housing
  - Warm housing
  - Sensors and controllers used to maintain and modify environment

Feeding and Nutrition

- Nutritional needs for energy, protein, fat, minerals and vitamins specified by NRC (National Research Council).
  - http://www.nap.edu/catalog.php?record_id=13298
- Energy provided by carbohydrates and fat in diet.
  - Needs vary with age, activity, production and temperature
- Protein
  - Provides animal with essential amino acids
  - Needs vary with age, sex and pregnancy
- Fat
  - Provides energy
  - Provides essential fatty acids
- Minerals
- Vitamins
- Water (not a nutrient but plenty of fresh, clean water is essential)

Chapter Resource

Balanced Diet Considerations

- Normal size and weight for pigs of a given age, sex, genetics and productive stage.

- Expected ages of puberty, length of estrous cycle, gestation length and litter size.

- Expected rate of gain for growing and finishing pigs.

- Expected average daily feed consumption and feed efficiencies for all stages of production.

- Producers should recognize symptoms of nutritional deficiencies or toxicities.

Feeding Practice Groupings

- Boars and gestating females
- Farrowing sows
- Litters
- Nursery pigs
- Growing and finishing pigs

Herd Health Management Concerns

- Items of concern that could introduce diseases.

    o **Biosecurity** issues (procedures intended to protect animals against disease or harmful biological agents).

        ✓ Herd additions
        ✓ Isolation of new animals
        ✓ People movement
        ✓ Pig movement

    o Other biosecurity risks

        ✓ Disease transmission on-farm
        ✓ Vermin control
        ✓ Vehicles
        ✓ Water and feed

Surveillance, Diagnosis, Treatment and Control of Disease

- Herd health program in place

- Vaccination program necessary

- External and internal parasite control program

- Regular observation of sick animals to prevent problems

Chapter Resource

- Immediate and proper diagnosis and treatment of sick animals

Pork Quality Assurance Plus

- Participation in the Pork Quality Assurance Plus (**PQA-Plus**)

- A management education program run by Pork Checkoff

- Emphasizes good management practices in the handling and use of animal health products.

- Encourages producers to review their approach to the herd's health problems.

- A workable, credible and affordable solution to assure food safety and animal care to help meet the needs of customers including restaurants, food retailers and consumers.

- Certification of a producer assures consumers they are purchasing the highest quality, safest product possible while caring for animal well-being.

- PQA Plus is a three-step program involving:

    1. Individual certification through education
    2. Farm site assessment
    3. Third-party verification to continuously analyze the success of the program.

- Website: http://www.pork.org/pqa-plus-certification/

Summary

Common breeds of swine in the United States include Berkshire, Chester White, Duroc, Hampshire, Landrace, Poland China, Spot and Yorkshire. Husbandry and management systems vary depending on the location, stage of production and type of production scheme. Some producers use environmentally controlled buildings for the animals. Farrowing systems help in the prevention of crushing of piglets and must comply with laws and regulations. Proper manure management is necessary to protect soil, water and air quality and must comply with laws and regulations. Nutrition of swine varies with age, activity, production and temperature. Safe, efficient management requires proper handling, good records, proper transportation and a herd health program. The Pork Quality Assurance Plus program is a means for producers to assure food safety and animal care to help meet the needs of customers including restaurants, food retailers and consumers.

Additional Resources

Taylor, R.E. and T.G. Field. 2010. Scientific farm animal production: Introduction to animal science. 10th ed. Upper Saddle River, NJ: Prentice Hall.

Breeds of Swine
http://www.ansi.okstate.edu/breeds/swine/

Pork Checkoff
http://www.pork.org/certification/11/pqaplus.aspx#.UnQya_mTh48

Swine Production and Management Home Study Course
http://extension.psu.edu/courses/swine/basic-production/introduction-to-swine-production

Assessment

Take assessment online here: http://tinyurl.com/AnSci-Swine
Download and print the assessment by scanning this QR code or by going to this URL: http://www.tagmydoc.com/Ch45AnSci

46 Sheep

Major Concept

Sheep are small ruminants that serve humankind by providing meat and fiber.

Objectives

- List the main characteristics of sheep
- List six predominant breeds/types of sheep
- Name the major concerns and problems of owning sheep
- List the components of a good health program for sheep

Key Terms

- Carrying capacity
- Ketosis
- Mutton
- Physiological condition
- Ram power

Chapter Resource

Complementary *full color* illustrations, photos, charts and graphs are available by scanning this QR code or by following this URL: http://www.tagmydoc.com/AS46 This digital resource will enhance your understanding of the chapter concepts.

Characteristics of Sheep

- Ruminant

    o Forage or roughage eater

    o Similar to cattle

- Gestation period: 147-150 days

    o Normal lambing rate

    o 1-4 lambs/year

    o 110 to 140% lamb crop

- Normal breeding season is fall but can breed out of season
    - Decreasing day length
- Produce wool
    - Shorn once or twice annually
- Mature breeding weight:
    - Ewes – 85 - 250 lbs
    - Rams – 160 - 350 lbs

Predominant Breeds

- Meat type breeds
    - Dorset
    - Suffolk
    - Hampshire
- Wool breeds
    - Columbia
    - Corriedale
    - Merino
- Hair breeds
    - Barbados Blackbelly
    - St. Croix
    - Katahdin
- Multiple births
    - Finnsheep
    - Polypay
    - Booroola Merino
    - Romanov

Management of Sheep

- Management concerns because sheep are prone to be:
    - Fragile
    - Disease prone
    - Susceptible to internal and external parasites

Chapter Resource

- Can be pasture managed in large flocks or confined

- Can use low quality roughage by-products or high-quality forages and grains

- Lambs are finished for slaughter before 1 year of age.

- Sheep older than 1 year are **mutton**.

- Problems

    o Predators

        ✓ Dogs, coyotes, eagles, hawks

    o Parasites

        ✓ Internal and external

    o Market

        ✓ Restricted outlet with few slaughter plants
        ✓ Lack of understanding and interest in sheep

Potential

- Forage from row crops plus wheat pasture forages can be used as effectively for sheep production as for cattle if managed properly

- Use of grazing land unsuitable for cultivation

    o Especially important with crop areas prone to erosion.

Chapter Resource

- Economic Outlook

    o Overall, lamb prices have been higher than cattle prices and more stable.

    o Capital necessary to purchase ewes to enter the sheep industry is less than what is required for cattle.

- Marketing Limitation

    o Mutton (lamb older than one year) is not a valuable product.

    o Less flexibility in finishing and marketing lambs compared to marketing and finishing beef cattle.

- Cost of Producing Lamb and/or Wool
  - Costs
    - ✓ Ewe maintenance ~ $50 - $100 per head annually
    - ✓ **Carrying capacity** (the number of people, other living organisms, or crops that a region can support without environmental degradation) of pasture land for sheep relative to cattle is about 5 ewe units to 1 cow unit.
    - ✓ Costs are comparable on a 5:1 ratio cost/lb. of gain in the feedlot is like cattle since their efficiency of gain (feed consumed/lb. of live unit gain) is about 7:1.

Production

- Meat production in sheep is more efficient primarily due to twinning.
- Frequency of twinning in sheep ~ 40-60%.
- Many lambs (~ 30-40%) can be marketed as fat lambs at about 5-7 months of age.
- Fat lambs average 100-130 lbs at marketing condition or finish.
- Normally require 45-60 days on high grain or lush pasture diets.
- Some are marketed directly off the ewe if creep fed.

Chapter Resource

- Wool production and value is another asset characteristic only of sheep.
- Wool represents less than 20% of the gross return.
- Lamb is responsible for 80% or more of the gross return in fine wool breeds.
- More interest in hair breeds for heat tolerance.
- Lamb production must be emphasized and improved more in the future.

Future Needs

- Carcass quality of lambs must be improved to eliminate the waste of fat lambs.
- New marketing outlets and more stable markets need development.
- Numbers must be increased for packers to stay busy and continue operation.

- Supply would be more stable.
- Wool must be de-emphasized and realigned for its true economic worth.
- Government subsidies may not always exist.
- Confinement management of sheep offers a bright future with a minimum of capital investment and could be developed.

Management Needs

- Breeding and Reproduction
    - Adequate **ram power** (number of rams/number of ewes).
    - Under pasture breeding conditions, one mature ram can service 30-35 ewes in a 60-day breeding period.
- Use breeds to maximize lamb production.
    - Many ewes can be rebred following weaning of their lambs and therefore lambing interval may be reduced to less than one year.

- Traditional system is one lambing each year.
- Accelerated lambing system.
    - Five lambing's in three years
    - Developed at Cornell University

Nutrition of Sheep

- Nutritional state is critical during all phases of production.
    - Most important prior to and during lambing.
    - Adequate energy consumption is the primary factor of concern.
    - Vitamin and mineral nutrition is probably more critical in a confinement system of production.
        - ✓ Vitamin A, D and E.

✓ Minerals include Na, P, Mg, Ca, Fe, I and Se.

- Protein and energy needs change with each **physiological condition** (age, sex and pregnancy). Specific needs must be recognized and met.

- Other nutritional concerns:

  o Need to more efficiently use available forage

  o Improved grazing systems

  o Use of available by-products

Chapter Resource

Lamb Management Activities

- Navel dip
- Identify
- Nurse
- Dock
- Castrate

Components of Health Program

- Enterotoxemia and tetanus management

- Proper vaccination for known disease threats

- Ketosis (pregnancy toxemia) in pregnant ewes

  o Controlled through diet

- External parasites

  o Spray or dip

- Internal parasites

  o Drenching program

  o Proper pasture rotation

- Prevention of tetanus in newborn lambs

  o Iodine solution to navel

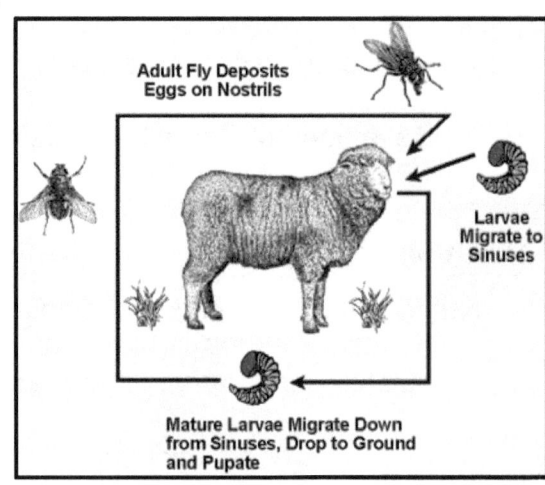

Nasal Bot Fly

Facilities

- Must be available to isolate and properly care for the ewes and lambs.
- Management and labor make the difference.
- Attention during lambing saves lambs.

Chapter Resource

Wool

- Lambs of wool breeds
    - Tails docked at earliest possible date.
    - Shearing is normally practiced once annually at the beginning of warm weather.
    - Freshly shorn sheep are very susceptible to cold, wet weather.
    - Prevention of fly infestation during shearing should be practiced.
    - Costs for custom shearing can vary widely depending on location per sheep.

Summary

Sheep are ruminants and forage or roughage eaters with gestation periods of 147 to 150 days. They can produce 1 to 4 lambs per year. The predominant breed types are meat, wool, hair and multiple births. Management concerns are fragility, disease prone and susceptible to internal and external parasites. Problems with sheep are predators, parasites and marketing. Marketing represents a limitation since mutton (lamb older than one year) is not a valuable product. Meat production in sheep is more efficient primarily due to twinning. Future needs of the industry are carcass quality of lambs to be improved to eliminate the waste of fat lambs; wool must be de-emphasized and realigned for its true economic worth.

Additional Resources

Ensminger, M.E. and R.O. Parker. 1986. Sheep and goat science. Danville, IL: Interstate Printers.

Taylor, R.E. and T.G. Field. 2010. Scientific farm animal production: Introduction to animal science. 10th ed. Upper Saddle River, NJ: Prentice Hall.

American Sheep Industry Association
www.sheepusa.org/

Guide to Raising Healthy Sheep
http://learningstore.uwex.edu/assets/pdfs/A3858-02.pdf

Sheep and Goats
http://www.nal.usda.gov/animals-and-livestock/sheep-and-goats

Sheep Production as an Alternative Enterprise
http://www.aces.edu/pubs/docs/A/ANR-0678/

Assessment

 Take assessment online here: http://tinyurl.com/AnSci-Sheep
Download and print the assessment by scanning this QR code or by going to this URL: http://www.tagmydoc.com/Ch46AnSci

47 Goats

Major Concept

Goats are small ruminants that serve humankind by providing meat and fiber.

Objectives

- Identify the predominant breeds/types of goats
- Outline the proper care and management of goats
- Identify some components of feeding and health care program for goats

Key Terms

- Balling gun
- Browse
- Carrying capacity
- Emasculator
- Forbs

Chapter Resource

Complementary *full color* illustrations, photos, charts and graphs are available by scanning this QR code or by following this URL: http://www.tagmydoc.com/AS47 This digital resource will enhance your understanding of the chapter concepts.

Overview of Goats

- Goats associated with people dating back 12,000 years.

- Provided milk, meat, skins for clothing and mohair.

- Consumption of goat's milk and goat meat in the U.S. is increasing.

- Production of mohair is an important industry in Texas.

- Thrive on **browse** (brushy plants) and **forbs** (broad leaf plants).

BOER

- Eat grass extensively, depending on forage species and stage of maturity.

- Less likely than sheep to suffer from internal parasites.

Predominant Breeds/Types

- Dairy Goats
  - Used largely to produce milk and to a lesser extent for meat.
- Major dairy goat breeds
  - Toggenburg
    - ✓ Medium in size, vigorous and capable of high milk productivity.
    - ✓ U.S. record for milk production for one lactation period is 6,650 lbs.

  - Saanen
    - ✓ Large all-white breed.
    - ✓ World record milk production for one lactation is 5,929 lbs.
  - Alpine
    - ✓ Large and somewhat rangy.
    - ✓ Record lactation is 6,415 lbs.

Chapter Resource

  - Nubian
    - ✓ Tall goat breed with long, wide, pendulous ears and a Roman nose.

- ✓ Better meat producer.
- ✓ Low milk production but has a higher milk fat content (7.4%).
- ✓ Considered a dual-purpose breed.

    o LaMancha

    - ✓ Smaller and differs from all other breeds in having no external ears (gopher ears) or extremely short ears (elf or cookie ears).
    - ✓ Milk production is comparatively lower.

    o Oberhasli

    - ✓ Medium-sized dairy goat breed, of Swiss origin, with usually solid red or black colors.
    - ✓ Production records of up to 3,300 lbs of milk have been reported.

- Angora Goat

    o Used mainly to produce mohair; also for brush clearance and meat production.

- Meat Goat

    o Boer is a popular breed.

    o Also, LaMancha and Nubian breeds could be considered.

Chapter Resource

- Cashmere Goat

    o Noted for soft cashmere fibers used in producing high-quality clothing.

- Pygmy Goat

    o Used as a laboratory ruminant animal and pets in the U.S.

    o Important disease-resistant meat and milk producer in West Africa and other countries.

Management of Goats

- Simple housing is adequate in mild weather areas.
- Goats do best when they are outside on pasture or in an area where they can exercise freely.

- Use of several small pastures increases grazing efficiency and reduces the risk of parasitic infection.

- Properly constructed fences are needed to keep goats away from neighbors, trees and gardens.

- Equipment needed for care: tattoo set, hoof shears and hoof knife, grooming brush, an **emasculator** (for castration) and a **balling gun** for administering large pills for dosing animals.

- Goats' feet should be properly trimmed to prevent deformities and footrot.

    o Footrot should be treated with formaldehyde, copper sulfate or iodine.

Balling Gun

- All kids should be identified by a tattoo in the ear, except for LaMancha and Pygmies, where identification is in the tail web.

- Male kids not acceptable for breeding should be castrated.

    o Caution must be taken to prevent infection, gangrene and other complications.

- Udders of dairy goats should have strong fore and rear attachments and the two teats should be well spaced.

Feeding and Health Program

- Goats are ruminants and can digest roughage effectively.

    o Does and bucks not actively breeding can perform satisfactorily on ample browse, good pasture, good quality grass and legume hay.

    o Short grass and poor-quality hay will require supplemental concentrates for feeding.

Chapter Resource

    o Lactating goats require roughage (hay) or another source of long fiber to prevent scouring.

    o Grains such as corn, oats, barley and milo may be fed whole because goats crack grains by chewing.

    o Deficiencies may exist in some areas so it is best to provide a mixture of trace-mineralized salt and bonemeal.

- o Young, growing and lactating goats require more protein than bucks or dry does.
- o Dairy goat kids must be allowed to nurse for the first three days (for colostrum), then can be removed and given a milk replacer.
- o By 2-3 weeks kids should be encouraged to eat solid foods such as leafy legume hay and palatable fresh concentrates such as rolled oats.
  - ✓ Small kids need concentrates until their rumens are sufficiently developed to digest enough roughage to meet their nutritional needs.

- Diseases and parasites
  - o Major goat diseases
    - ✓ Johne's disease
    - ✓ Caseous lymphadenitis
    - ✓ Caprine pleuropneumonia
    - ✓ Ecthyma
    - ✓ Enterotoxemia
    - ✓ Goat pox
    - ✓ Herpesvirus
    - ✓ *Pasteurella hemolytica*
    - ✓ Tetanus
    - ✓ Viral leukoencephalomyelitis

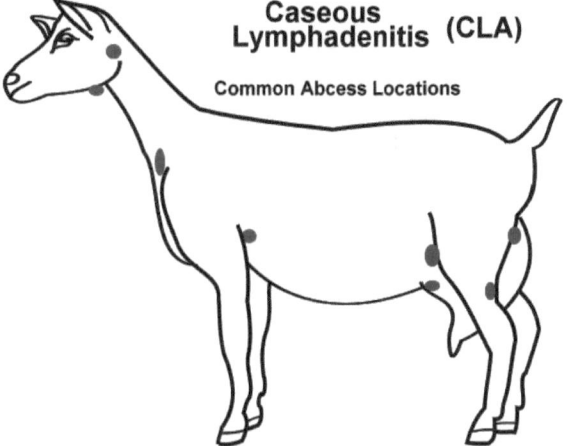

Production

- Considerations of special importance in managing dairy goats are: care at time of breeding, kidding and feeding.

- Female goats come into heat at intervals of 15 to 18 days until she becomes pregnant.

- Most goats are seasonal breeders – Normal season is September, October and November with kids being born in February, March and April.

- Normal goat lactations are 7 to 10 months.

- With staggered breeding, the kidding and milking seasons can be extended.

- Housing goats in the dark for several hours each day in spring and summer months (to simulate the onset of shorter days) causes some to come into estrus earlier than usual.

- Artificial additional light in the goat barn may delay estrus in the fall.

• One service is all that is needed to obtain pregnancy, but it is wise to delay breeding for a day after the goat first shows signs of heat.

- Does stays in heat from 1 to 3 days but the optimum period of standing heat may last only a few hours at the end of estrus.

• Careful records should be kept showing breeding dates and the bucks used.

- Detailed breeding records are essential for knowing ancestors which helps in the selection of superior males to produce especially desirable offspring.

• Delivery of twins is common and delivery of triplets occur occasionally among mature does.

- Kids weigh 5 to 8 lbs at birth (depending on breed).

- Singles may be heavier than doubles or triplets.

- Males are usually heavier than females.

Chapter Resource

• At kidding, a doe should be placed in a clean pen, well bedded with clean straw or shavings.

- Observed and not disturbed unless assistance is necessary as indicated by excessive straining for 3 hours or more.

- If assistance is needed, a qualified attendant should pull when contractions occur.

• As soon as the kid arrives, its mouth and nostrils should be wiped clean of membranes and mucus.

- Kids must be dried and warmed immediately in cold weather.

- Kids should be encouraged to nurse as soon as possible to help the newborn stay warm.

• Difficulty in kidding may be caused by a low selenium (Se) supply, diseases and internal parasites.

Summary

Goats have been associated with people dating back 12,000 years. Throughout history they provided milk, meat, skins for clothing and mohair. Predominant types include: dairy goats, Angora goat, meat goats, cashmere goat and pygmy goat. Goats do best when they are outside on pasture or in an area where they can exercise freely. Use of several small pastures increases grazing efficiency and reduces the risk of parasitic infection. Goats are ruminants and can digest roughage effectively. Grains such as corn, oats, barley and milo may be fed whole because goats crack grains by chewing. Considerations of special importance in managing dairy goats are care at time of breeding, kidding and feeding. Delivery of twins is common. Difficulty in kidding may be caused by a low selenium supply, diseases and internal parasites.

Additional Resources

Ensminger, M.E. and R.O. Parker. 1986 Sheep and goat science. Danville, IL: Interstate Printers.

Taylor, R.E. and T.G. Field. 2010. Scientific farm animal production: Introduction to animal science. 10th ed. Upper Saddle River, NJ: Prentice Hall.

Dairy Goats
http://www.agmrc.org/commodities_products/livestock/goats/dairy-goats/

Goat Program
http://www.ansci.cornell.edu/4H/meatgoats/meatgoatfs2.htm

Sheep, Goats and Small Ruminants
http://www.nal.usda.gov/animals-and-livestock/sheep-and-goats

Assessment

Take assessment online here: http://tinyurl.com/AnSci-Goats
Download and print the assessment by scanning this QR code or by going to this URL: http://www.tagmydoc.com/Ch47AnSci

48 Horses

Major Concept

Horses of various breeds and types of are used for work and recreational activities.

Objectives

- Identify various breeds of horses
- Classify breeds as draft, light horses or ponies
- List five general uses of horses
- Name five gaits of horses
- Explain the various factors used in judging horses
- Name the parts of a horse
- Identify colors of horses
- Identify signs of a healthy horse
- List good management practices for feeding horses and maintaining their health
- Identify basic horsemanship procedures
- List points to observe when judging horses
- Name types of horse shows or competitions

Key Terms

- Breed
- Canter
- Conformation
- Draft
- Floating
- Gait
- Gallop
- Hand
- Pace
- Rack
- Sound
- Tack
- Trot
- Walk

Chapter Resource

 Complementary *full color* illustrations, photos, charts and graphs are available by scanning this QR code or by following this URL: http://www.tagmydoc.com/AS48 This digital resource will enhance your understanding of the chapter concepts.

Breed Definition and Color

- A group of animals of the same species that share common traits.

- Colors – color classification varies based upon breeds.

- Bay – any shade of brown with a black mane and tail and often black on the legs.
- Chestnut – generally a darker brown with a mane and tail of the same color.
- Sorrel – a lighter brown with mane and tail of same color.
- Grey – usually has dark skin. Coat can be any shade of grey with either dapples (areas of spotted grey) or flea-bit (very small spots of darker or reddish gray all over the body).
- Black – black coat and skin must be black.
- White – very rare. Coat must be white, as well as skin. Generally referred to as an albino if it has pink eyes.
- Palomino – coat is the color of a new-minted penny, with white mane and tail.
- Buckskin – yellow color coat with black mane, tail and legs.
- Dun – can be various shades of yellow, always has a dorsal stripe.
- Roan – can be strawberry (red & white) or blue (black & white).

Gaits

- Walk – slowest of gaits, 4 beats.
- Trot – medium gait – legs work in diagonal pairs producing 2 beats.

Walk

Trot

- Pace – medium gait – legs work in side pairs producing 2 beats.
- Canter – faster gait legs work in 3 steps (ex. left hind, opposite diagonal pair, right front).

- Gallop – fastest gait – legs move independently (ex. left hind, right hind, left front, right front).

- Rack – a gait in which the horse appears to be walking behind and trotting in front.

Major Horse Categories, Breeds and Characteristics

- Four general categories of horses:

    1. Draft
    2. Light
    3. Gaited
    4. Ponies

- **Draft** - These are large horses that usually stand taller than 16 hands at the withers (1 hand = 4 inches). These horses are very muscular and large boned, and generally used for work such as pulling heavy loads.

    o Belgian – Characterized by its usual color of chestnut with flaxen mane and tail.

    o Shire – Largest of all draft breeds.

    o Clydesdale – A fairly light draft breed made famous by Budweiser.

    o Percheron – Usually black or gray, often crossed with lighter horses to make heavy riding horses.

- Light horses – This is what most people envision when a horse is mentioned. These are, by far, the most numerous in terms of population and breed. Light horses are generally used for riding and light carriage work.

    o Light horses, generally have only three gaits: walk, trot, canter.

    o Quarter Horse – The most numerous breed in the U.S. made famous by cowboys and rodeos. Characterized by heavily muscled frame and angular face.

- Appaloosa – A versatile breed developed by the Nez Percé Indians from Spanish stock. The breed was nearly wiped out by the U.S. Calvary during the Indian Wars. Characterized by various patterns of spots.

- Arabian – The most ancient of all purebred light horses. This breed originated in the deserts of Arabia. Characterized by a finely chiseled head, dished face, long arching neck and high tail carriage. The Arabian is known for its endurance and density of bone which makes it a popular cross on many breeds.

- American Paint – This is a stockhorse type (like Quarter Horses) that is characterized by its loud, splotched markings. These markings can be in any color.

- Standard bred – This is a light racing horse that is known for its speed at a trot or a pace. These horses race pulling sulkies (light weight cart for one person).

- Thoroughbred – A popular breed originating in Europe. Thoroughbreds are used in a variety of sports although most commonly associated with racing.

- Morgan – An American breed developed in New England. Characterized by flashy gaits, hardiness and versatility. The tail is attached high and carried gracefully and straight. Morgans appear to be a strong powerful horse and the breed is well known as an easy keeper.

- Gaited horses are also light riding horses; but in addition to walk, trot and canter, these horses also rack and do other fancy footwork.

    - American Saddlebred – This breed is characterized by its flashy movements both front and rear and a very high head carriage.

    - Tennessee Walking Horse – Developed by plantation owners in the south, the walking horse's movements are characterized by high steps in the front and long reaching strides in the rear.

    - Missouri Fox Trotter – A gaited horses used mostly for pleasure.

    - Paso Fino – A small Spanish horse originating in the Caribbean. The Paso Fino is notable for its Paso Gait - a rapid rhythmic gait that is so smooth people are seen riding them carrying full glasses of water.

- Ponies are small horses whose height cannot exceed 14.2 hands.

    - Shetland – Developed in the Shetland Isles, this is perhaps the most well-known of Pony breeds. Shetland Ponies are extremely handy and, when well-maintained, can live for 35 years.

- Welsh Ponies – This is a very light pony, originating in Wales and known for its refinement. Very popular in shows and the Welsh pony is very versatile.
- Hackney Pony – Noted for its flashy movements, the Hackney is most often seen pulling a carriage.

Chapter Resource

Uses of Horses

- Pleasure
- Breeding
- Working stock
- Show
- Sport

Selecting and Judging Horses

- Requires knowledge and information
- Breeds
- Conformation/structure and parts of a horse
- Soundness
- Movement

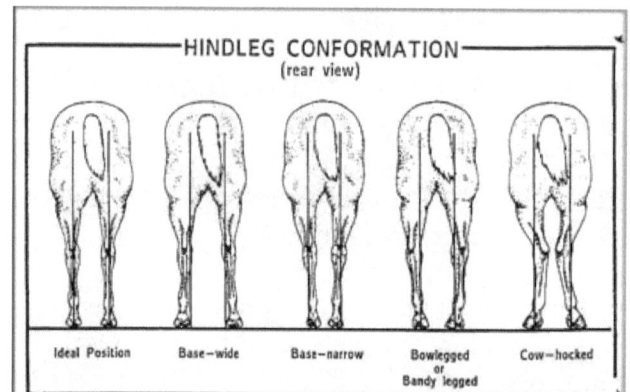

Parts of a Horse and Their Ideal

- Poll is the bony prominence lying between the ears. Except for the ears, it is the highest point on the horse's body when it is standing with its head up.

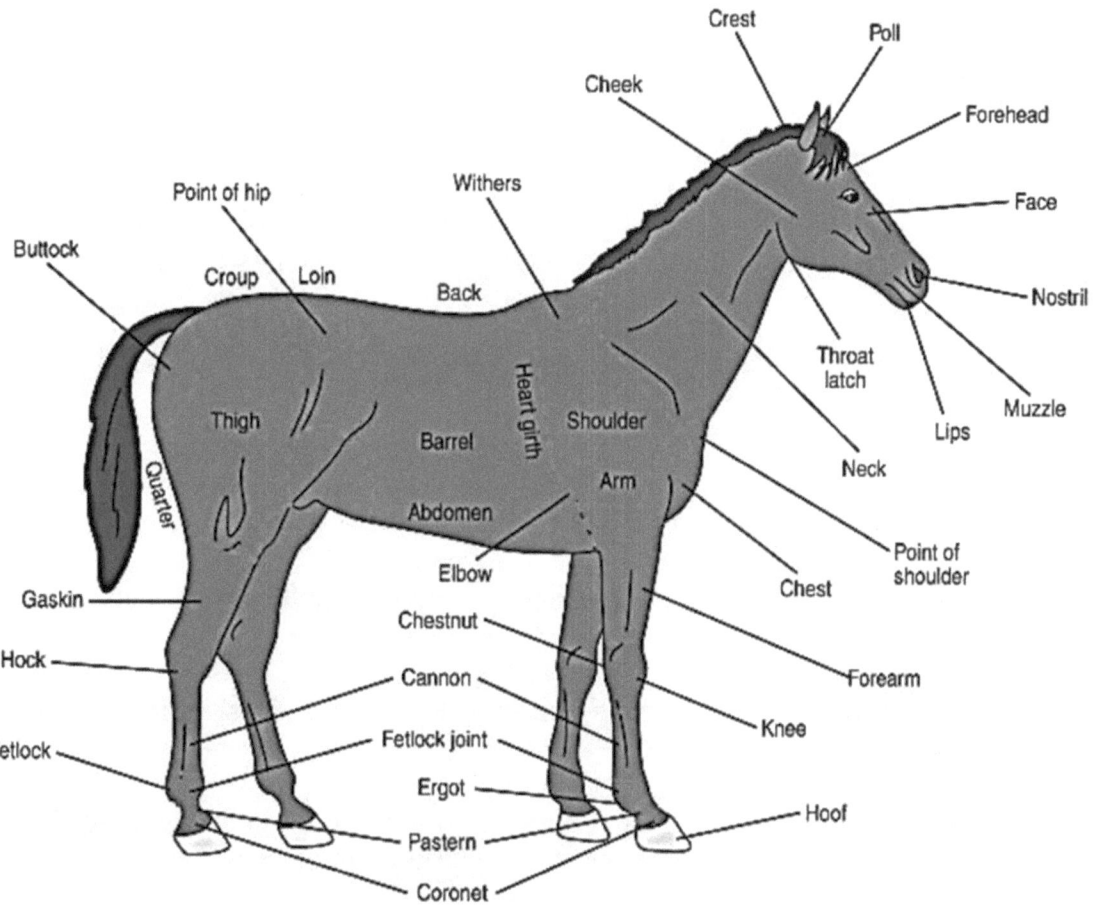

- Crest is the curved top line of the neck. It should be moderately lean in mares but inclined to be fuller in stallions.

- Forehead should be broad, full and flat.

- Nostrils should be capable of wide dilation to permit the maximum inhalation of air, yet be rather fine.

- Head should taper to a small muzzle, the lips should be firm and the lower lip should not have the tendency to sag.

- Point of shoulder is a hard, bony prominence surrounded by heavy muscle masses.

- Breast is a muscle mass between the forelegs, covering the front of the chest.

- o An ideal chest is deep and contains the space necessary for vital organs. A narrow chest can lead to interference with the front legs. Chest muscles should be well developed and form an inverted "V". The prominence of chest muscling depends on the breed.

- Forearm should be well muscled: it extends from the elbow to the knee.

- Knee is the joint between the forearm and the cannon bone.

Chapter Resource

- Coronet is the band around the top of the hoof from which the hoof wall grows.

- Hoof refers to the horny wall and the sole of the foot. The foot includes the horny structure and the pedal bones and navicular bones, as well as other connective tissue.

- Pastern extends from the fetlock to the top of the hoof.

- Flexor tendons run from the knee to the fetlock and can be seen prominently lying behind the cannon bone, when it runs parallel to the cannon bone it constitutes the desired "flat bone."

- Fetlock is the joint between the cannon bone and the pastern. The fetlock joint should be large and clean.

- Cannon bone lies between the knee and fetlock joint and is visible from the front of the leg. It should be straight.

- Hock is the joint between the gaskin and the cannon bone, in the rear leg. The bony protuberance at the back of the hock is called the point of hock.

- Gaskin is the region between the stifle and the hock.

- Stifle is the joint at the end of the thigh corresponding to the human knee.

- Flank is the area below the loin, between the last rib and the massive muscles of the thigh.

- Loin or coupling is the short area joining the back to the powerful muscular croup (rump).

- Croup (rump) lies between the loin and the tail. When one is looking from the side or back, it is the highest point of the hindquarters.

- Back extends from the base of the withers to where the last rib is attached.

- Withers are the prominent ridge where the neck and the back join. At the withers, powerful muscles of the neck and shoulders attach to the elongated spines of the second to sixth thoracic vertebrae. The height of a horse is measured vertically from the withers to the ground, because the withers are the horse's highest constant point.

- Neck should be fine at the throat latch to allow the horse ease of flexing. Neck should blend smoothly into the withers and the shoulders and not appear to emerge between the front legs. Lightweight horses should have reasonably long necks for good appearance and proper balance.

- Shoulders should be overlain with lean, flat muscle and blend well into the withers.

- Barrel should be narrower at the shoulders and widen at the point of coupling (loins).

- Girth is the point that a horse should be measured to determine the heart girth which can be used to determine the horses weight.

- Elbow is a bony prominence lying against the chest at the beginning of the forearm.

- Hindquarters give power to the horse, should be well muscled when viewed from the side and rear.

Conformation

- The physical structure of a horse's or pony's body.

- Dictates athletic ability and ability to stay **sound** (one who has no lameness or illness).

- Used when judging horses.

- Some points covered in Parts of a Horse and Their Ideal.

Chapter Resource

- Set of legs important to ability.

- View from the front

    o Ideal leg set: Vertical line from point of shoulder should fall in center of knee, cannon, pastern and foot.

    o Problems:

        ✓ Toes out

- ✓ Bow-legged
- ✓ Narrow-chested
- ✓ Base narrow (stands close)
- ✓ Knock-kneed
- ✓ Pigeon-toed

The Basics of Horse Conformation- Purdue YDAE: https://www.youtube.com/watch?v=nQINNTIeNJM

- View of back legs from side
  - Ideal leg set: Vertical line from point of buttock should touch the rear edge of cannon from hock to fetlock and meet the ground behind the heel.
  - Problems:
    - ✓ Stands under
    - ✓ Camped-out
    - ✓ Leg too straight

Chapter Resource

- View of front legs from side
  - Ideal leg set: Vertical line from shoulder should fall through elbow and center of foot.
  - Problems:
    - ✓ Camped-under
    - ✓ Camped-out
    - ✓ Buck-Kneed
    - ✓ Calf-kneed

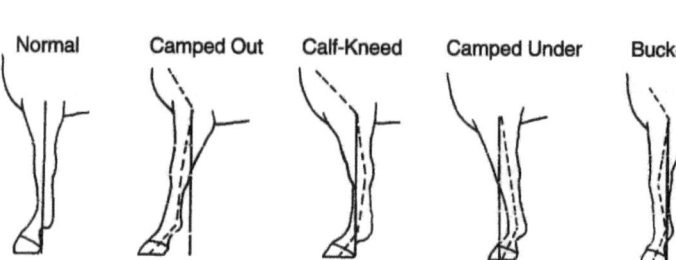

- View of legs from the back
  - Ideal legs set: Vertical line from point of buttock should fall in center of hock, pastern and foot.
  - Problems:
    - ✓ Stands wide
    - ✓ Stands close
    - ✓ Bow-legged
    - ✓ Cow-hocked

- Important points to observe when judging horse for purchase or for judging competition
    - Head
    - Femininity or masculinity
    - Chest capacity
    - Set of legs
    - Width of croup and width through rear quarters
    - Style and beauty
    - Balance and symmetry
    - Neck
    - Shoulders
    - Topline
    - Coupling
    - Rear flank
    - Arm, forearm and gaskin
    - Legs, feet and pasterns
    - Quality
    - Breed type
    - Soundness
    - At walk
    - At trot
    - At canter
- Years of practice and knowledge of the parts of a horse, conformation, gaits and soundness are required to become a judge.

Feeding Horses

- Feeds should:
    - Furnish horses with a daily supply of nutrients in the correct amounts.
    - Supply palatable, easily obtained feedstuffs.
    - Provide feedstuffs economical for the conditions.
- Feeds for horses include five groups:
    1. Roughages
    2. Concentrates
    3. Protein supplements
    4. Minerals
    5. Vitamins

- By nature, horses consume forage.
- Under natural conditions, they spend several hours a day grazing.

- Basing rations on adequate amounts of good-quality roughage minimizes digestive disturbances such as colic.

- Supplementing hay or pasture with the correct amount of the right concentrates will meet all requirements for energy, protein, minerals and vitamins.

- Since individual horses vary considerably in their nutrient requirements, feeding horses is both an art and a science.

- All horses require nutrients to maintain body weight and to support digestive and metabolic functions. In some cases they need additional nutrients for growth, work, reproduction or lactation.

Chapter Resource

- Nutrient requirements for horses expressed in two ways:

    1. Daily nutrient requirements
    2. Nutrient concentration in the feed (This may be expressed on an as-fed basis or on a dry-matter basis.)

- Most horses receive their daily ration in two parts: roughage (hay, or pasture) and concentrates.

- Concentrate portion contains grain and may include a protein supplement, minerals and vitamins.

- Concentrate may also include bran, cane molasses, and/or dehydrated alfalfa.

- Horse owner must decide:

    o How much and what kind of roughage to feed.

    o Correct concentrate mixture and the amount needed to supply the nutrients not present in adequate amounts in the roughage.

Feeding Guidelines

- Feed only quality feeds.

- Feed balanced rations.

- Feed half the weight of the ration as quality hay.

- Feed higher protein and mineral rations to growing horses and lactating mares.

- Feed legume hay to young, growing horses; lactating mares; and out-of-condition horses.

- Use non-legume hays for adult horses.

- Regulate the hay-to-grain ratio to control condition in adult horses.

Chapter Resource

- Feed salt separately, free-choice.

- Feed calcium and phosphorus free-choice.

- Keep teeth functional. Horses 5 years and older should be checked annually by a veterinarian to see if their teeth need **floating** (filing).

- See that stabled horses get exercise – they will eat better, digest food better, and be less prone to colic.

- Feed according to the individuality of horse. Some horses are hard keepers and need more feed per unit of body weight.

- Feed by weight, not volume. A gallon of two different grains may vary in nutrient content.

3 qt Plastic – 3 lbs   3 qt Metal – 3.7   4 qt Metal – 5.5 lbs   6 qt Metal – 7 lbs   8 qt Bucket – 10 lbs

- Minimize the use of finely ground feedstuffs in a prepared ration. If a ration is ground fine, horses will be reluctant to eat it and the chances of colic will increase.

- Offer plenty of good water, no colder than 45°F. Free-choice water is best. Horses should be watered at least twice daily.

- Change feeds gradually. When changing from a low-density (low-grain), high-fiber ration to one of increased density, change gradually over a period of a week or more.

- Start horses on feed slowly. Horses on pasture should be started on dry feed gradually. Start this on pasture if practical, and gradually increase the feed to the desired amount in a week to 10 days.

- Do not feed grain to tired or hot horses until they have cooled and rested, preferably for 1 or 2 hours. Instead, feed hay while they rest in their blankets or out of drafts.

- Feed before work. Hungry horses should finish eating at least an hour before hard work.

- Feed all confined horses at least twice daily. If horses are working hard and consuming a lot of grain, three times is mandatory.

- When feeding hay, give half the hay allowance at night, when horses have more time to eat and digest it.

- Founder and colic are nutritional diseases of special concern for horse owners.

Signs of Good Health

- Body condition/weight (fatness or thinness)

- Hair coat – Shiny, glossy.

- Hoof growth – Normal growth rate, smooth and uncracked.

- Eyes – Bright, fully open, clear, without discharge.

- Normal hydration

- Normal feces and urine – Firm fecal balls, wheat colored clear urine.

- Healthy pink mucous membranes of gums and lips.

- Normal heart rate – 32 to 48 beats per minute.

- Normal respiration rate – 8 to 16 breaths per minute.

- Normal body temperature – 99.5 to 101.5°F.

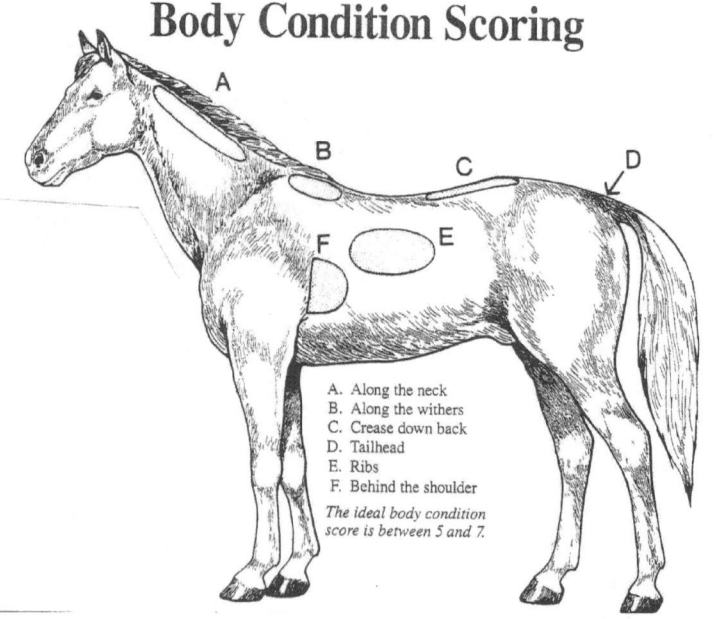

Body Condition Scoring

A. Along the neck
B. Along the withers
C. Crease down back
D. Tailhead
E. Ribs
F. Behind the shoulder

The ideal body condition score is between 5 and 7.

Horse Health Program

- Variety of diseases and parasites can affect the horse.

- A good health management program is the best prevention.

- Shelter from wind and weather with trees and a shed or barn is adequate in most climates.

Chapter Resource

- To reduce chance of injury, a safe environment that is free of hazards such as nails, barbed wire, broken fences, glass windows, and unsecured pesticides should be provided.

- Adequate clean water should be provided at least two times a day; free access to water is best.

- A routine schedule of feeding and exercise should be maintained. Sudden changes in feeds, feeding schedule, or work/activity can cause lameness, colic and muscle problems. Regular exercise, either free-choice or regulated, is important in maintaining athletic horses.

- Horses are natural nibblers. They can be fed once a day but will be more efficient (digest more of their feed) if fed two or three times daily.

- Horses should be fed at least 1.5 to 2.5% of their body weight per day in hay or pasture. Hay (forage) helps prevent intestinal problems and abnormal behavior (vices) caused by lack of fiber and boredom.

- Commercial concentrate feed mixtures should be used if necessary to supply the nutrients needed. The concentrate should be selected to complement the hay or pasture composition. Feed should be fed based on weight, not volume.

- Hays and feeds should be free of dust and mold.

- When necessary, feeds should be changed gradually over a 10 to 14-day period.

- Floating (filing) a horse's molars to decrease the sharp points that interfere with normal chewing is often needed. Regular dental checkups and floating will prevent mouth problems.

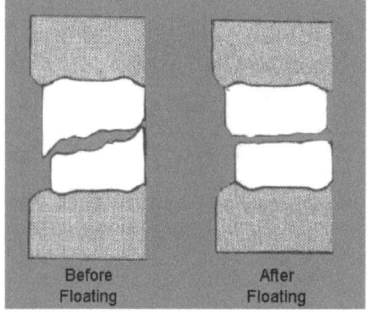

- Horses should be dewormed regularly, and parasite load should be assessed with occasional fecal floatation tests and treated accordingly.

- All horses should be on a regular vaccination schedule that includes tetanus.

- Regular hoof care is important. Feet may need to be trimmed periodically if hoof growth exceeds wear, but it is not usually necessary to keep shoes on horses that are not in training or being ridden or driven on rough terrains.

- When questions arise or problems develop, owners should seek help from a veterinarian.

Chapter Resource

Tack

- Equipment used for riding or driving horses.
    - Bridle
    - Saddle (stock, hunt-jumping, gaited, dressage, Western, English)
    - Harnesses

Basic Horsemanship

- Includes knowing how to:
    - Halter
    - Lead
    - Bridle
    - Saddle
    - Mount
    - Ride
    - Dismount

- Differences for English, Western, Hunt-jump, Gaited or Dressage horsemanship

Horse Shows and Competition Include

- Hunter
- Jumper
- Saddle
- Harness
- Western
- English
- Breeds
- Dressage
- Rodeo
- Cutting
- Polo
- Fox hunting
- Driving
- Gymkhanas
- Distance riding
- Racing
- Draft horse demonstrations/competitions

Summary

Breed is defined as a group of animals of the same species that share common traits. Color classification can vary based upon breeds. Gaits include walk, trot, pace, canter, gallop and rack. Major horse categories are draft, light horses, gaited horses and ponies. Ways in which horses are used include pleasure, breeding, working stock, show and sport. Selecting and judging horses requires knowledge and information, breeds, conformation/structure and parts of a horse, soundness and movement. Conformation dictates athletic ability and ability to stay sound, used when judging horses and set of legs important to ability. Feeds should furnish horses with a daily supply of nutrients in the correct amounts, supply palatable, easily obtained feedstuffs and provide feedstuffs economical for the conditions. Feeds for horses include five groups: roughages, concentrates, protein supplements, minerals and vitamins. Nutrient requirements for horses are expressed in two ways: Daily nutrient requirements and nutrient concentration in the feed. Know the feeding guidelines for a horse and implement a horse health program.

Additional Resources

Parker, R. 2013. Equine science. 4th ed. Clifton Park, NY: Delmar Cengage Learning.

Horse Extension
http://www1.extension.umn.edu/agriculture/horse/

USDA National Agricultural Library - Horses
https://www.nal.usda.gov/horses

Assessment

Take assessment online here: http://tinyurl.com/AnSci-Horses
Download and print the assessment by scanning this QR code or by going to this URL: http://www.tagmydoc.com/Ch48AnSci

49 Poultry

Major Concept

Rapid growth and vertical integration characterize the poultry industry.

Objectives

- List and explain the vertical integrated structure of the modern poultry industry
- Identify and describe common meat and layer breeds of chickens
- Identify and describe the most common meat breed of turkey
- List the steps involved in egg processing, egg packaging and egg labeling
- Identify the steps of poultry meat processing
- List the requirements of packaging labels

Key Terms

- Breeder
- Evisceration
- Feed mill
- Growout farms
- Hatchery
- Incubation
- Layer breed
- Marketing
- Meat breed
- Microbes
- Picking
- Plant code
- Primary breeder
- Processing plant
- Satellite farms
- Specialty designer eggs
- Variety
- Vertical integration

Chapter Resource

 Complementary *full color* illustrations, photos, charts and graphs are available by scanning this QR code or by following this URL: http://www.tagmydoc.com/AS49 This digital resource will enhance your understanding of the chapter concepts.

Vertically Integrated Structure of the Modern Poultry Industry

- Many segments of the poultry industry exist.
  - Segments are all interlinked and many times owned by the same company.
    - ✓ This type of ownership is called vertical integration.
- **Vertical Integration**

- A style of business management that allows for maximum control of the products produced.

    ✓ Much like a ladder concept as each segment relies on the segment below it to supply it with what it needs, while that segment must produce the product the next segment needs.

- A hierarchy of needs that are met within one company.

- Allows the poultry industry to develop its product efficiently and to produce a high-quality product.

* Structure – Consists of eight steps
    1. Primary breeders
    2. Feed mill
    3. Breeders
    4. Hatchery
    5. Growout farms
    6. Processing plants
    7. Further processing
    8. Transportation and marketing

* **Primary Breeders**

    - Responsibility is to develop and reproduce strains of chicken that meet the requirements of chicken producer/processing companies.

    - Through development and reproduction, they aim to achieve desirable characteristics such as abundant white meat and efficient feed conversion.

    - Breeder chicks with the appropriate mix of desirable characteristics are then sold to integrated chicken firms.

* **Feed Mills**

    - Chicken companies own feed mills that convert raw materials into finished feed according to very specific formulas developed by poultry nutritionists.

    - They produce about four to five different formulas to feed all the different nutrition stages of chickens.

* **Breeders**

    - Operated by contract growers who raise the breeder chicks to adult birds.

- o Breeding hens and roosters are kept under tight biosecurity on breeder farms to produce fertile hatching eggs.

- o Offspring of breeder parents will then be raised to become broilers for the market.

- **Hatchery**

  - o Specialized facility designed to hatch fertile eggs received from breeder farms.

  - o Fertile eggs are placed in incubators and carefully monitored to ensure that correct temperature and humidity levels are maintained throughout the entire **incubation** (to keep eggs in the proper conditions for development) period.

  - o Towards the end of incubation, the eggs are placed in hatching trays where the chicks hatch out by pecking their way through the large end of the egg.

- **Growout Farms**

  - o Newly hatched chicks are transported to growout farms where independent farmers raise them to market weight under contract with the company.

  - o Company provides the chicks, feed, and any necessary pharmaceuticals.

  - o Farmer provides the growout house, water, bedding ("litter"), electricity and his own management skill.

  - o Chickens reach market weight in six or seven weeks and are collected to be taken to the processing plant.

Chapter Resource

- **Processing Plants**

  - o Processing plant harvests the birds by humane standards and are inspected by the USDA for any disease or defects.

  - o Carcasses are then chilled in ice-cold water to limit the growth of bacteria.

  - o Following chilling, they are packaged for distribution or cut into parts.

- **Further Processing**

  - o Specialized operations or plants that receive whole chicken or cut-up parts and perform a variety of further processing steps.

- ✓ Steps include cooking, breading or marinating.

Chapter Resource

- **Transportation and Marketing**
  - ○ Chicken products are transported in refrigerated trucks from the processing and further processing plants to grocery stores, restaurants and other customers.

Value to the Economy

- Vertical integration within the poultry industry gives producers greater control over the production of quality products that successfully meet consumer wants and needs.

- Shown to be more cost effective for the company as well.

Common Breeds of Meat and Laying Chickens

- Chicken industry has well over 50 breeds that are recognized by the American Poultry Association.
  - ○ These breeds are broken down by classifications and use. The use of these animals would be layers, meat and dual use.

- Meat Breeds

- Not really breeds but hybrid varieties or combinations of many different breeds.
  - ○ Varieties are developed for specific characteristics:
    - ✓ Grow faster and larger
    - ✓ Larger breast meat yield
    - ✓ More efficient feed conversion
    - ✓ More disease resistance

- Varieties used by broiler producing companies that commercially produce broilers sold in supermarkets.
    - ✓ Weakness of these varieties
    - ✓ Do not lay as many eggs as the layer breeds.

- Specific Variety Used in Industry
  - ○ Cornish Cross

- ✓ White Cornish x White Plymouth Rock
- ✓ Fast growth allows them to reach 4-5 lbs in 6 weeks and 6-10 lbs in 8-12 weeks.

  o White Cornish

  - ✓ Part of the English Class
  - ✓ Very broad and meaty body

  o White Plymouth Rock

  - ✓ Part of the American Class
  - ✓ Tend to be docile and a good dual-purpose breed.

Chapter Resource

- Layers Breeds

  o Genetically selected for high egg productivity.

  o Tend to be small bodied so they are undesirable for meat production.

  o Small bodies allow the bird to put more nutrients toward egg production instead of body size.

  o Divided into two types

  1. Lay white or brown eggs
  2. Chicken breeds with white ear lobes lay white eggs, whereas chickens with red ear lobes lay brown eggs.

Specific Breeds Used in Industry

- White Leghorns

  o Part of the Mediterranean Class

  o Very good layer of white eggs

  o Basis of commercial egg industry

- Rhode Island Red

  o Part of the American Class

  o Lay brown eggs

Rhode Island Red

- Production-bred strains lay very well

Common Meat Breed Used in the Commercial Turkey Industry

- Currently eight breeds of turkeys that are recognized by the American Poultry Association.

- Several breeds exist that are not officially recognized as a breed but these are the varieties that are commercially used by the industry.

  - These breeds are predominately used for meat.

- **Meat Breed**

  - Broad Breasted White

    - Commercially the most widely-used breed of domesticated turkey.
    - Shorter breast bones and legs than "standard" turkeys.
    - Unable to breed naturally and require artificial insemination (AI).
    - Produce more breast meat and their pin feathers are less visible when the carcass is dressed due to their white color.

Processing Eggs

- Refer to Chapter 7 Milk and Eggs for information on processing eggs.

Poultry Processing

- Steps of poultry meat processing

  - Shackling

    - Poultry meat processing is initiated by hanging, or shackling, the birds to a processing line.
    - Birds are transferred from coops or transport cages to a dark room where they are hung upside down from shackles attached to an automated line.

  - Stunning

- ✓ Electrical stunning delivers a current through a water bath to deem the bird unconscious.
- ✓ Controlled environment stunning is another alternative stunning process. Birds are immersed in an approved gas or mixture of gases (i.e., CO_2) to displace oxygen and render the bird unconscious.

o Bleeding

Chapter Resource

- ✓ After stunning, the birds are passed through an automated knife that makes an incision on the neck to cut the jugular vein.
- ✓ With the carcass hanging upside down and the jugular vein cut, most of the blood in the carcass will exit.

o Scalding

- ✓ Scalding loosens the feathers to facilitate their removal.
- ✓ Carcasses are submerged into the scalder that contains water heated to 150°F.
- ✓ This high-water temperature serves to loosen the connection of feathers to the skin.

o Picking

- ✓ Picking is a term that refers to feather removal.
- ✓ Picker is an automated machine that contains rubber finger-like projections that rotate in a circular motion to remove feathers without damaging the carcass.

o Removal of feet, head, neck and oil glands

- ✓ Feet are removed at the knee joints.
- ✓ Head is cut and removed.
- ✓ Neck is cut with machine and esophagus is exposed.

o **Evisceration** refers to the removal of internal organs.

- ✓ Inedible viscera consist of the spleen, esophagus, lungs, intestines and reproductive organs.
- ✓ Intestines (viscera) are federally inspected for signs of disease or other problems.

- ✓ Identified disease or other problems results in the removal, or condemnation, of the carcass from the processing line.
- ✓ Edible viscera, or giblets, consist of the heart, liver and gizzard.
- ✓ Giblets are packaged in the carcass or sold separately.

Chapter Resource

- o Washing the carcass
    - ✓ Carcasses are cleaned for microbial and visible concerns. When processing chicken, microbial bacteria such as *E. Coli* and *Salmonella* are analyzed.
- o Chilling
    - ✓ Carcass temperature must be reduced to prevent microbial growth.
    - ✓ The USDA specifies the amount of chilling for specific bird sizes:
        1. 4 lb broiler: 40°F within 4 hours
        2. 4-8 lb broiler: 40°F within 6 hours
        3. Greater than 8 lb broiler: 40°F within 8 hours
    - ✓ Submerging the carcass in an ice (chilled water) bath is the most common method of carcass chilling.
    - ✓ Carcass can also be chilled by air chilling.
    - ✓ Air chilling occurs by passing cold air over the carcass. This is a more expensive process but some consumers are willing to pay more for air chilling.
- o Cut-up and deboning

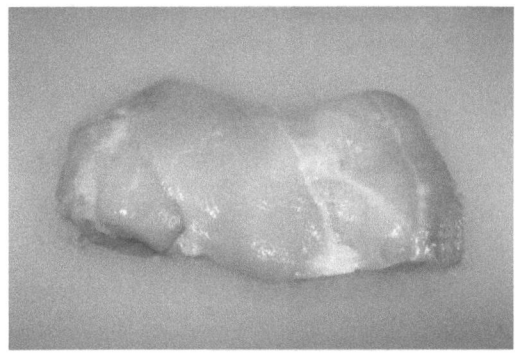

  - ✓ On average, 75-80% of the live animal weight is retained in the carcass.
  - ✓ This amount of live weight retained in the carcass is known as the dressing percentage.
  - ✓ Carcass can be sold whole, or individual components of carcass can be cut-up for individual sale.
  - ✓ Cut-up often includes removal of the breast, thigh, drumstick and wings.
  - ✓ Deboning refers to the removal of bone from the cut-up meat.
  - ✓ Breasts and thighs are commonly deboned.

Further Processing

- Whole carcass or cut-up and deboned pieces may be further processed for added value.

- Further processing may include forming, curing, smoking and cooking of products.

- Forming product requires a change in particle size and often includes the addition of ingredients to add flavor.

- Forming product also requires the use of a mold to obtain desired shape.

- Formed products include hot dogs, chicken nuggets or sausage.

- Curing involves the addition of preservatives, often nitrates, to the meat to improve flavor and product shelf-life.

- Smoking also acts as a preservative while providing additional flavor to the product.

- Some product may be prepared in a form that is edible without additional preparation and is known as "ready-to-eat."

Storage

- Poultry meat should be refrigerated at approximately 30°F. Keeping the poultry below 40°F reduces the risk of microbial growth.

- Refrigeration and freezing does not kill all microbes; some will survive.

Requirements of a Packaging Label

- Sell By

    o Product dating not required

    o Preferred by stores for dating

    o Date used for quality assurance

    o Food can still be used if chicken is frozen and the sell by date is expired.

- Plant Code

    - The **plant code** traces the facility that produced the product and tracks the product in case of a recall.

- Grade of chicken

    - Grade A – Best quality (plump, bruise free, no broken bones)

- Nutritional facts

    - Serving sizes

    - Fat, sugar, sodium and carbohydrate inclusion.

- "Keep Refrigerated"

- Following label regulations

    - Labeling claims must be valid

        ✓ For example: if labeled "Organic", the bird must have been raised organically and certified organic.

    - Some labels are misleading

        ✓ For example: "No hormones or steroids" – It is illegal to use hormones or steroids in poultry production.

Chapter Resource

Summary

Many segments of the poultry industry exist. Segments are all interlinked and many times owned by the same company. This type of ownership is called vertical integration which is a style of business management that allows for maximum control of the products produced. The structure consists of primary breeders, feed mill, breeders, hatchery, grow-out farms, processing plants, further processing and transportation and marketing, all having specific responsibilities. Vertical integration within the poultry industry gives producers greater control over the production of quality products that successfully meet consumer wants and needs. The chicken industry has well over fifty breeds that are recognized by the American Poultry Association; however, the industry uses hybrid varieties selected for the production of meat or eggs. In-line egg processing occurs at the same location as the egg production facility. Whereas, off-line egg processing occurs separate from the egg production facility. Broad Breasted White turkeys are commercially the most widely-used breed of domesticated turkey. Poultry meat processing requires specific steps to follow. Further processing may include forming, curing, smoking, and cooking of products. Poultry meat requires storage the proper temperature to reduce the risk of microbial growth. Labels on poultry products must follow government guidelines that require specific information.

Additional Resources

Taylor, R.E. and T.G. Field. 2010. Scientific farm animal production: Introduction to animal science. 10th ed. Upper Saddle River, NJ: Prentice Hall.

American Poultry Association
http://www.amerpoultryassn.com/

Choosing a Chicken Breed: Egg, Meat, or Exhibition
https://www.extension.purdue.edu/extmedia/as/as-518.pdf

National Chicken Council
http://www.nationalchickencouncil.com

The Pacific Egg and Poultry Association (PEPA)
http://www.pacificegg.org

The United Egg Producers (UEP)
http://www.unitedegg.org/

The United States Department of Agriculture (USDA) Egg Grading Manual
https://www.ams.usda.gov/sites/default/files/media/Egg%20Grading%20Manual.pdf

Assessment

Take assessment online here: http://tinyurl.com/AnSci-Poultry
Download and print the assessment by scanning this QR code or by going to this URL: http://www.tagmydoc.com/Ch49AnSci

50 Aquaculture

Major Concept

Aquaculture provides a healthy food source to a growing human population and does so through the efficient use of resources.

Objectives

- Define aquaculture
- Name five aquatic species raised for food
- Compare the energy requirements of aquatic animals to traditional livestock
- Identify the general nutrients needed by aquatic animals
- List four methods of feeding aquatic animals
- Identify four physical forms of feed used

Key Terms

- Aquaculture
- Carnivores
- Carotenoids
- Glycogen
- Herbivores
- Monogastrics
- Omnivores
- Poikilothermic

Chapter Resource

Complementary *full color* illustrations, photos, charts and graphs are available by scanning this QR code or by following this URL:
http://www.tagmydoc.com/AS50 This digital resource will enhance your understanding of the chapter concepts.

Aquatic Species

- **Aquaculture** – the raising of aquatic animals or the cultivation of aquatic plants for food.

    o Raised in ponds, raceways, tanks and cages in ponds, lakes, rivers or oceans.

    o Holds great hope for the future of feeding an increasing population.

Leopard Frog

- Common animal species include:

    - Alligators
    - Carp
    - Channel catfish
    - Clams
    - Crayfish
    - Frogs
    - Prawns
    - Salmon
    - Shrimp
    - Tilapia
    - Trout

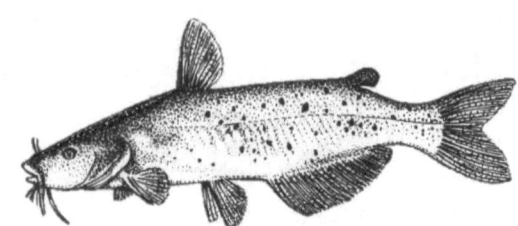

- Catfish and trout most common fish raised in the U.S.

- Other species are gaining in popularity.

Characteristics of Fish

- **Poikilothermic** - having body temperature that varies with the environment. (Mammals and birds are homoeothermic.)

- In the wild, source of feed is natural foods.

- Species are classified as warm water, cold water, fresh or salt water.

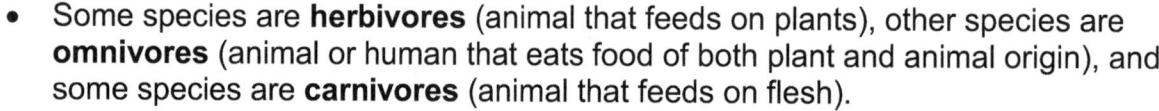

- Some species are **herbivores** (animal that feeds on plants), other species are **omnivores** (animal or human that eats food of both plant and animal origin), and some species are **carnivores** (animal that feeds on flesh).

- Fish animals are **monogastrics** (have a simple single-chambered stomach).

Dietary Protein Needs

- Higher than mammals.

- Essential amino acids required in diet:

- Trout 40-46% crude protein
- Carp 31-38% crude protein
- Catfish 32-36% crude protein
- Salmon 40% crude protein
- Shrimp 40% crude protein

Dietary Carbohydrate Needs

- Used poorly

- Digestibility only 20 to 40%

- Excess stored as **glycogen** – a substance deposited in bodily tissues as a store of carbohydrates.

Dietary Lipid Needs

- Fats major source of energy.

- Metabolism different than land animals.

- Essential fatty acid requirement.

Chapter Resource

Dietary Energy Needs

- Lower maintenance energy than mammals and birds because:
  - Energy not needed to regulate body temperature
  - Energy needed to overcome gravity is less because body density near that of habitat
  - Less energy expended searching for food

- Efficient feed conversion compared to land animals.

Vitamin Needs

- Same vitamins as land animals.
  - Vitamin C required in diet (functions in connective tissue formation).

- **Carotenoids** (natural pigments found in plants and animals) are used for pigmentation; for example, the pink flesh of some fish.

Mineral Needs

- Same minerals as land animals.

- Many obtained from minerals in the water.

    o Calcium (Ca) is met by absorption in water

    o Phosphorus (P) dietary requirement

Chapter Resource

Toxins

- Sensitive to toxins

- Mycotoxins and especially aflatoxin

Feeding Programs

- Producers encourage the growth of natural foods in growing environment.

- Producers provide feed in one of these forms:

    o Pellets

    o Crumbles

    o Meals

    o Flakes

Pellets

- Depending on type, the feed maybe sinking or floating.

Feed Formulation and Processing

- Correct feed size for species and age of animal.

- Stable nutrients, resistant to leaching, since feeding in water environment.

- Proper physical properties for species being fed.

Feeding Methods

- By hand
- Machine blowers
- Automatic feeders
- Demand feeders (self-feeder)

Typical Diet Components

- Fish meal (or other meat meal)
- Soybean meal (or another oilseed meal)
- Corn/grain
- Middlings
- Fat source
- Trace minerals and vitamin premix
- Dicalcium phosphate
- Ascorbic acid

Nutrient	Minimum	Maximum
Crude protein (%)	28	28
Crude fat (%)	–	6
Crude fiber (%)	–	7
Available lysine[1] (%)	1.43	–
Available methionine and cystine[2] (%)	0.65	–
Available phosphorus (%)	0.30	0.40
Digestible energy (kcal/g)	2.6	2.8
Yellow pigments[3] (ppm)	–	7

Sample feed for Channel Catfish

Aquaculture Activities

- Five general activities performed to produce a product include:
    1. Hatchery
    2. Grow-out
    3. Harvesting
    4. Processing
    5. Marketing

Chapter Resource

Summary

Aquaculture provides a healthy food source to a growing human population and does so through the efficient use of resources. It is defined as raising of aquatic animals or the cultivation of aquatic plants for food. Aquaculture animal species are poikilothermic.

Their body temperature varies with the environment. Species can be herbivores, omnivores or carnivores. Each species has unique requirements for dietary protein, carbohydrates, fats, vitamins and mineral. Processed and properly formulated feeds supply needed nutrients or in some cases natural food from the environment are used. Aquatic animals require less energy than land animals to maintain body functions so more energy is available for growth. Five general activities performed to produce a product include hatchery, grow-out, harvesting, processing and marketing.

Additional Resources

Parker, R. 2012. Aquaculture science. 3rd ed. Clifton Park, NY: Delmar Cengage.

Aquaculture Extension
https://appliedecology.cals.ncsu.edu/extension/aquaculture/

Understanding Fish Nutrition, Feeds and Feeding. Virginia Cooperative Extension
http://pubs.ext.vt.edu/420/420-256/420-256.html

What is Aquaculture?
http://www.nmfs.noaa.gov/aquaculture/what_is_aquaculture.html

Assessment

Take assessment online here: http://tinyurl.com/AnSci-Aquaculture
Download and print the assessment by scanning this QR code or by going to this URL: http://www.tagmydoc.com/Ch50AnSci

51 Companion Animals

Major Concept

Recognize that companion animals or pets represent a recent addition to animal science.

Objectives

- Identify four types of companion animals
- List four breeds of dogs and four breeds of cats
- Identify how companion animals connect to agriculture
- List two reasons people keep pets
- Name some major expenditures associated with companion animals
- Identify improvements in breed programs as more is known about DNA

Key Terms

- Companion animal
- DNA certification
- Human-developed
- Natural breeds
- Pet
- Reptiles/amphibians
- Spontaneous mutations

Chapter Resource

 Complementary *full color* illustrations, photos, charts and graphs are available by scanning this QR code or by following this URL: http://www.tagmydoc.com/AS51 This digital resource will enhance your understanding of the chapter concepts.

Place of Pet and Companion Species in U.S.

- Many households own more than one type of **pet** (a domestic or tamed animal or bird kept for companionship or pleasure and treated with care and affection).

- Types of pets and facts about pet ownership:

  o Dogs – 36.5% of U.S. households as of 2012 (www.avma.org)

  o At least one cat – 30.4% of U.S. households

- Birds – 3.1% of households
- Horses – 1.5 % of households
- Fish
- Small mammals
- **Reptiles/Amphibians** (a cold-blooded vertebrate animal of a class that comprises the frogs, toads, newts and salamanders)

Chapter Resource

Who Owns Pets?

- Families with children
- People over 65 and living alone
- Higher household income
- Homeowners
- More people in a house

- Factors driving the overall pet population include:
    - Move of the baby boom generation into the most common pet owning years
    - Increased resources for pet ownership due to smaller family size
    - Security and companionship of a pet against factors of modern life
    - Surrogate bonding with the natural world through pets
    - Increases ecological awareness
    - Dog's traditional role as protector

How Companion Animals Connect to Agriculture

- Agriculture provides most of the raw ingredients for food and other products used in animal upkeep.
- Veterinary costs can be substantial.

- Millions of dollars spent annually for pet products.
- Cat food sales are increasing more rapidly than dog food sales.
- Pet industry supports several related industries like the pet food and the pet supplies industries.

Chapter Resource

Pet and Companion Animals Defined

- In the classic definition, **pets** are different from livestock because they are kept for pleasure rather than utility.
  - The modern term "**companion animal**" is used to indicate that a human is frequently in the company of an animal.
  - Animals that are not companion animals.
    - ✓ Animals are not necessarily domestic species.
    - ✓ Tamed specimens of wild animals.
  - Pets can be classified as:
    - ✓ Ornamental or exotic pets
    - ✓ Status symbols
    - ✓ Playthings
    - ✓ Hobby animals
    - ✓ Work, helper or service animal
    - ✓ Companion animals
    - ✓ Treated as a member of the family; includes dogs, cats, small mammals or birds.

- Other animals included as pets or small animals used by humans:
  - Laboratory species (rats and mice)
  - Hedgehogs
  - Hamsters
  - Rats
  - Guinea pigs
  - Parakeets
  - Other birds
  - Reptiles
  - Amphibian

Historical Perspective of Dogs and Cats

- Gray wolf, *Canis lupis*, is the ancestor of the first domestic animal, the dog.

    o About 12,000 B.C.

- African bush cat, *Felis lybica* ancestor to domestic cats

    o About 6,000 years ago

Chapter Resource

Genetics and Breeding Programs

- Similar to other species

- Species selected for breeding

- Undesirable genetic diseases

- Majority of breeders only rudimentary knowledge of breeding and genetics.

- Perhaps rapidly advancing state of DNA technology knowledge about genetics will improve.

    o Dog genome project

    o Inexpensive technology already available to positively identify the parentage of all dogs.

    o In the future, possible to compare the genetic similarity between the individuals of a proposed mating.

    o Dogs screened for all genetic diseases.

- DNA Certification

    o The Board of Directors of the American Kennel Club approved a voluntary DNA Certification Program in 1998.

    o Program offers Certificates of DNA Analysis.

    o Voluntary DNA Certification Program.

    o Also provides a means of identification for individual animals.

Development of Breeds

- **Natural Breeds**
    - Selected by human preference or regional diversity (Examples: Abyssinian, Birman, Burmese and Maine Coon).

- Human-developed breeds

    Chapter Resource

    - Created by crossbreeding and subsequent selection to fix type (Examples: Bombay and the Himalayan).

- **Spontaneous mutations**
    - New breeds that showcase a mutation [Examples: American Curl (ears curled back), American Bobtail (short tailed), Munchkin (short legs)].

Breeds of Dogs

- Between 400 to 450 breeds of dogs.

- Primary dog registry of the U.S. is the American Kennel Club, (AKC), http://www.akc.org/ which classifies purebred dogs into seven categories.

    1. Sporting
    2. Hounds
    3. Working
    4. Terriers
    5. Toys
    6. Nonsporting
    7. Herding
    8. Miscellaneous

- Ten Most Popular Breeds of Dogs

    1. Labrador Retriever
    2. Golden Retriever
    3. German Shepherd
    4. Dachshund
    5. Beagle
    6. Poodle
    7. Yorkshire Terrier
    8. Chihuahua
    9. Boxer
    10. Shi Tzu

Breeds of Cats

- Less numerous than dogs.
- Fewer uses overall for cats.

- Approximately 50 breeds of cats.

- Breeds of cats and photos at these websites:

    o Cat Fanciers' Association homepage
       http://www.cfainc.org

    o Animal Planet: Cat Breed Directory
       http://animal.discovery.com/guides/cats/selector/selector.jsp

- Longhaired Breeds of Cats

    o Balinese
    o Kashmir
    o Ragdoll
    o Birman
    o Maine Coon
    o Somali
    o Cymric
    o Norwegian Forest
    o Tiffany
    o Himalayan
    o Persian
    o Turkish Angora
    o Javanese

- Shorthaired Breeds of Cats

    o Abyssinian
    o American
    o American Wirehair
    o Bombay
    o British
    o Burmese
    o Chartreux
    o Colorpoint
    o Cornish Rex
    o Devon Rex
    o Egyptian Mau
    o Exotic Shorthair
    o Havana Brown
    o Japanese Bobtail
    o Javanese
    o Korat
    o Malayan
    o Manx
    o Ocicat
    o Oriental Shorthair
    o Russian Blue
    o Scottish Fold
    o Siamese
    o Singapura
    o Snowshoe
    o Sphynx (hairless)
    o Tonkinese

Reproductive Characteristics of Various Pets/Small Animals

- Reproductive traits are varied and very different from traditional livestock.

Species	Puberty	Cycle Length	Duration of Estrus	No. Young	Gestation
Dog	5-24 mo.	3 1/2-13 mo.	6-12 days	1-22	58-70 days
Cat	4-12 mo.	14-21 days	6-7 days	1-10	58-70 days
Rabbit	5-9 mo.; range 4-12 mo.	none	To 1 mo.	1-6	28-31 days
Rat	37-67 days	4-5 days	~12-18 hrs	2-20	20-22 days

Species	Puberty	Cycle Length	Duration of Estrus	No. Young	Gestation
Mouse	28-49 days	Usually 4-5 days	Few hrs	1-12	17-21 days
Guinea Pig	55-70 days	16 1/2 days	6-11 hrs	1-6	59-72 days
Hamster	4-6 wks.	4-5 days	12 hrs	1-12	14 days
Gerbil	9-12 wks	4-6 days	12-15 hrs	2-15	24-26 days

Species	Reproductive Cycle Type
Dog	Unseasonably mono-estrous
Cat	Induced ovulation, seasonally polyestrous and early fall
Rabbit	Induced ovulation; breed all year long; more or less; may show seasonal anestrus
Rat	Polyestrous all year
Mouse	Polyestrous all year
Guinea Pig	Polyestrous all year
Hamster	Polyestrous all year; few pregnancies in winter
Gerbil	Polyestrous

Chapter Resource

Nutrition of the Pet Species

- Good nutrition for a pet is responsibility of owner.

- Premixed feeds generally purchased with attention to information on the product label.

- Pet food industry is a large and growing industry.

- Major focus is production of specialized products.

- Industry is expanding to include more sales of dog and cat foods by veterinarians.

- Demand growing for higher quality, higher priced brands typically sold in this way.

- Life stage concept feed formulations and marketing.

- Diets marketed for specific types of performance.

Guaranteed Analysis:
- Crude Protein .... Min. 21 %
- Crude Fat .... Min. 13 %
- Crude Fiber .... Max. 12 %
- Moisture .... Max. 11 %
- Ash .... Max. 5.5 %
- Calcium .... Min. 0.5 %
- Phosphorus .... Min. 0.4 %
- Vitamin E .... Min. 400 IU/kg
- Ascorbic Acid* (Vitamin C) .. Min. 75 mg/kg

*Not recognized as an essential nutrient by the AAFCO Dog Food Nutrient Profiles.

Trends in Veterinary Expenditures

- Cost of veterinary care is increasing in all animal groups.

- Number of visits per cat increasing substantially.

- Specialty and exotic pet owners making increased use of veterinary services.

- More medical technology is available to pet owners.

Chapter Resource

Summary

Many households own more than one type of pet, such as dogs, cats, birds, horses, fish, small mammals and reptiles. Companion animals are connected to agriculture because agriculture provides most of the raw ingredients for food and other products used in animal upkeep. In the classic definition, pets are different from livestock because they are kept for pleasure rather than utility. The modern term "companion animal" is used to indicate that a human is frequently in the company of an animal. The development of breeds includes natural breeds, human-developed breeds and spontaneous mutations. Good nutrition for a pet is the responsibility of the owner.

Additional Resources

American Kennel Club
http://www.akc.org/

Animal Planet: Dogs
http://www.animalplanet.com/pets/dogs/

Cat Fanciers' Association homepage
http://www.cfainc.org

Humane Society – Companion Animals
http://www.humanesociety.org/about/departments/companion_animals.html

USDA National Agricultural Library - Companion Animals
http://awic.nal.usda.gov/companion-animals

Assessment

Take assessment online here: http://tinyurl.com/AnSci-CompanionAnimals
Download and print the assessment by scanning this QR code or by going to this URL: http://www.tagmydoc.com/Ch51AnSci

52 Animal Behavior

Major Concept

Animals exhibit various behaviors in relation to their surroundings, other animals and their physical needs.

Objectives

- List six general behaviors and give examples
- Give two examples of instincts
- Define ethology and anthropomorphism

Key Terms

- Abnormal behavior
- Agonistic behavior
- Allomimetic behavior
- Anthropomorphism
- Eliminative behavior
- Epimeletic behavior
- Ethology
- Grooming behavior
- Ingestive behavior
- Instincts
- Investigative behavior
- Mimicry
- Reactive behavior
- Sexual behavior

Chapter Resource

 Complementary *full color* illustrations, photos, charts and graphs are available by scanning this QR code or by following this URL: http://www.tagmydoc.com/AS52 This digital resource will enhance your understanding of the chapter concepts.

Definitions of Animal Behaviors

- **Ethology** is the study of animal behavior – group and individual actions which take place to allow them to live and function in their environment. It also consists of their learned abilities and inherited tendencies.

- **Instincts** are the reflexes and behavior patterns animals are born with.

- Ten major patterns of behavior include:

    1. Reactive
    2. Ingestive

3. Eliminative
4. Sexual
5. Care-giving and care-seeking
6. Agonistic
7. Mimicry
8. Investigative
9. Grooming
10. Sleep or rest

Care-giving and care-seeking

- o **Reactive behavior** is simply the animal reacting to its surroundings such as communicating and visual contact with the rest of the herd, a reflex to pain and discomfort or seeking shelter.

- o **Ingestive behavior** includes the mechanics of eating and chewing, obtaining food and water – the daily patterns of feeding.

    - ✓ The first feeding trait of all mammals is suckling. Farm animals usually ruminate after grazing for several hours. Rumination is the regurgitation of feed for chewing.

- o **Eliminative behavior** involves voiding of feces and urine.

    - ✓ Most large animals defecate while standing or walking and wherever the animal is.
    - ✓ Hogs prefer to defecate in specific areas of a pasture or pen.
    - ✓ Animals defecate and urinate more frequently when stressed or excited.

- o Sexual behavior observations are important for determining times for breeding.

    - ✓ **Sexual behavior** involves the courtship, mating and maternal behavior and is controlled by hormones but may be learned.
    - ✓ Different animals show different behavior when estrus (heat) occurs. Generally, when the female will allow the male to mount her, this indicates standing heat. Males also exhibit mating behaviors that the female responds to.

- o Caregiving and care-seeking behaviors are also called **epimeletic behavior**.

    - ✓ Animals seek attention and care from each other.

- o **Agonistic behavior** includes fighting or flight and other reactions involving conflict.

- ✓ Dominance and hierarchy in the herd is established through agonistic behavior.
- ✓ Males are more likely to fight than females.

Chapter Resource

- o **Mimicry** involves animals simply doing what the other animals in the herd or group are doing.
    - ✓ This is also called **allelomimetic behavior**.
    - ✓ Presence of other animals in the group provides companionship.
- o **Investigative behavior** is shown when animals explore or investigate a new environment or object.
    - ✓ Younger animals are more apt to be curious than older animals.
- o **Grooming behavior** is seen between animals or as they groom themselves.
    - ✓ They may roll in dry dirt or rub a part of their body when scratching.
- o Sleep or rest behavior allows the animal to restore its physical status.
    - ✓ Animal may be drowsy but wakeful.
- **Abnormal behaviors** vary with species.
    - o Occurs because some animals cannot adapt to their environment, boredom or poor nutrition

Anthropomorphism

- Attributing human characteristics to animals.
- It can be a problem, when observing animals, to use phrases and comments which do not reveal anything about what is actually happening.
    - o Observations like: "the bull is mad", "someone made him angry" etc. do not tell us about the behavior – they simply project human emotions to animals.
    - o This is not to say that animals have no feelings or emotions.
- Animals experience pain, hunger, fear and rage along with other emotions including stress and excitement.
    - o Comments like "the bull's head is lowered", "the eyes are wide and open", "he appears to be pawing the ground" are all valid observations to be made.

- o Guesses such as "the bull's territory has been invaded" and "the bull is being threatened by someone or something" are also valid guesses, rather than "someone made him angry" etc.

- o This is because the first two comments imply a cause for the behavior rather than just a sudden display of anger.

Summary
Animals exhibit ten general types of behaviors: reactive, ingestive, eliminative, sexual, care-giving and care-seeking, agonistic, mimicry, investigative, grooming and sleep or rest. Some behaviors are learned and others are instinctive. Knowing and understanding animal behaviors will minimize problems and increase success in handling of the animals.

Additional Resources
Taylor, R.E. and T.G. Field. 2010. Scientific farm animal production: Introduction to animal science. 10th ed. Upper Saddle River, NJ: Prentice Hall.

Animal Behavior - Khan Academy
http://tinyurl.com/qfgusbk

Behavioral Principles of Livestock Handling
http://www.grandin.com/references/new.corral.html

Dr. Temple Grandin's Web Page of Livestock Behaviour, Design of Facilities and Humane Slaughter
http://www.grandin.com/

Assessment

Take assessment online here: http://tinyurl.com/AnSci-AnimalBehavior
Download and print the assessment by scanning this QR code or by going to this URL: http://www.tagmydoc.com/Ch52AnSci

53 Issues in Animal Science

Major Concept

Major issues face the animal industries today.

Objectives

- Describe three issues facing each of these animal industries: dairy, beef, sheep, equine, poultry and aquaculture
- Compare animal welfare to animal rights
- Identify the concerns around biotechnology
- List three animal welfare concerns for each of the livestock species

Key Terms

- Animal rights
- Animal welfare
- CAFOs
- Environmentalism
- Federal land
- PSS – Porcine Stress Syndrome
- Wool Act

Chapter Resource

 Complementary *full color* illustrations, photos, charts and graphs are available by scanning this QR code or by following this URL: http://www.tagmydoc.com/AS53 This digital resource will enhance your understanding of the chapter concepts.

Beef Issues

- Food safety
- Integration and concentration
- Value-based marketing
- Endangered species
- Product perception
- Quality assurance
- Global markets
- Public lands
- Water quality and quantity
- Air quality
- Biotechnology
- Animal welfare
- Global warming/climate change

Dairy Issues

- Milk marketing
- Air quality and water quality and quantity
- Animal welfare
- Product perception
- Global markets
- Biotechnology
- Food safety
- Diet and health

Horse Issues

- Animal welfare
- Water quality and quantity
- Global markets
- Biotechnology

Poultry Issues

- Food safety
- Animal welfare
- Quality assurance

Chapter Resource

Sheep Issues

- Product perception
- **Wool Act** – The Wool Act of 1699 is an Act of the Parliament of England which attempted to heighten taxation and increase control over colonial trade and production.
- Public lands
- Predator control
- Quality assurance
- Global markets
- Animal welfare

Swine Issues

- Integration and concentration
- Food safety
- Diet and health
- Quality assurance
- Air quality and water quality and quantity
- Global markets
- Value-based marketing

Animal "Protection" Groups

- Many view animal production as exploitation.
- Worry about animal use and control.
- Promote **animal welfare** (physical and psychological well-being of animals) with extremes of **animal rights** (rights believed to belong to animals to live free from use in medical research, hunting, and other services to humans) and in some cases the liberation of animals from human control.

Animal Welfare Concerns: Beef

- Branding, castration and dehorning seen as cruel and unnecessary
- Worry about cancer eye and treatment of downer cows
- Promote humane slaughter
- Promote humane handling in feedlots and handling during transport

Animal Welfare Concerns: Swine

- Express concern over confinement systems, the use of farrowing crates, flooring systems.
- Objections to tail docking, teeth clipping, castration and early weaning.
- Halothane gene which indicates susceptibility to stress: **Porcine Stress Syndrome (PSS)**.
- Promote proper handling and transport.
- Issues with wild/feral pigs in some parts of the country.

Animal Welfare Concerns: Dairy

- Concerns over calf management, housing and stall systems
- Objections to castration, dehorning, branding, tail docking
- Concerns about the frequency of mastitis and lameness
- Welfare of downer cows

Animal Welfare Concerns: Veal

- Objection to white veal production

- Concerns of behavioral deprivation and flooring systems used for veal

Animal Welfare Concerns: Poultry

- Objections to housing/cages

- Concerns about behavioral problems due to production systems

- Objections to management practices such as forced molting, debeaking, toe trimming

- Concerns that poultry do not get exercise, not allowed to nest, take a dust bath and that birds become bored

Animal Welfare Concerns: Horses

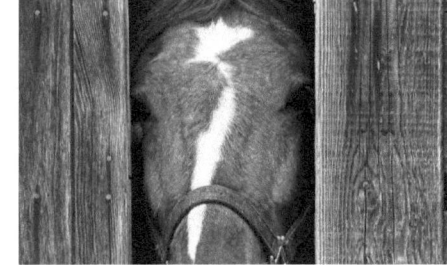

- Objections to some training techniques

- Concerns about proper housing, stalling and transporting

- Objections to racing as two-year-olds

- Slaughter of old horses and wild horses seems unacceptable to many but without this what to do with the animals

- Wild horses are protected and some adopted but in some areas, are a problem

- Concerns about sports injuries (racing), insurance fraud and the use of performance drugs

Model for Livestock Producers from Animal Protection Groups

- Not suffer pain or distress unless part of experiment or unless control of pain or distress would invalidate experiment

- Not be used repeatedly for invasive experiments

- Drugs causing paralysis while leaving animal without anesthesia

- Husbandry and housing fits nature of the animal

- Oversight of these principles provided by local committees thru protocol review and facility inspection

Animal Welfare Act

- For a full text of the Act visit

    o http://www.nal.usda.gov/awic/legislat/awicregs.htm

- For the Animal Welfare Information Center visit

    o http://www.nal.usda.gov/awic/

Chapter Resource

- Animal Rights and Liberation in the news

    o Some see animals as having the same "rights" as human

    o Humans and animals are not different – all life the same

Biotechnology

- Application of physical, chemical and engineering principles to biological systems

- Ranges from use of vaccines to manipulation of genome

- Potential to improve human life

- Consumer notions/perceptions vary

    o Where do the consumers stand today?

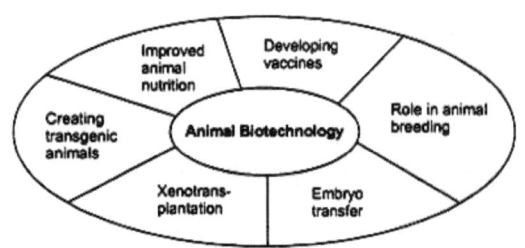

Public Impressions of Genetic Engineering

- It is science and technology.

- Often think of test-tube babies and other items from the news

- Combinations of plant/animal/human

- Negative/Frightened

- Consider the possibility of creating monsters, mutants and unknown and uncontrollable bacteria and viruses

- Most are familiar with the term DNA and some understanding of it

- Some recognize use in medicine and progress

- Many just "don't know."

Chapter Resource

Biotechnology Issues

- Offers hope but not "silver bullet."

- Often not simple

- Lag time from discovery to market is often a long period

- Always creating new moral and ethical issues

- Other new issues such as patents of the genetic code

- Safety concerns can be real or perceived

Environmental Issues

- Early ideas of U.S.: "land of milk and honey" endless

- "Greatest good for the greatest number."

- Henry David Thoreau and others in late 19th century.
    o Wilderness unique and some protection from human influence.

Environmentalism

- First, as a philosophy opposed to private use of public resources.

- Believe that biosphere preservation first priority: human needs secondary.

- Create recycling and restrictive regulations to use natural resources.

- o   Need to sacrifice standard of living to protect the environment.

Livestock Environmental Issues

- Waste management, water use and quality in large confinement units: **Confined Animal Feeding Operation (CAFOs)**.

- Federal lands belong to everyone.

- Air quality; for example, the "smell" created by large dairy operations and settling ponds.

- Endangered species competition with livestock.

- Livestock perceived contributions to global warming/climate change.

- Conversion of agricultural land to urban development.

Chapter Resource

Waste Management

- Main issues are manure and odor.

- Industry generally willing to change.

- Requires stewardship of livestock producers.

- New methods for handling wastes include composting and biogas generation.

Water Use and Quality

- Livestock seen as nonpoint source pollution from direct contamination and due to amount of use.

- Use of commercial fertilizers and manure can contribute to pollution.

- Proper planning of facilities and drainage for livestock operations.

- Requires proper pesticide use, storage and application.

Federal Lands

- Large amount in western states:

    - Nevada = 86%

    - Utah = 64%

    - Idaho = 63%

Chapter Resource

- Issues become who uses and who pays for use; for example, grazing of cattle and sheep.

Endangered Species

- Act of 1973

- Recovery plan needed for endangered species.

- Producers need to know how many are listed, how many have recovery plans, and how many de-listings are occurring.

- Fact vs. emotion often cloud the issue; for example, in the case of wolves.

Global Warming/Climate Change

- Climate change is more frequently used than global warming.

- Big effort to make it a political issue as well.

- Maybe yes; maybe no.

- May be effect of all human action.

- Consider to be caused by increased CO_2 and CH_4 levels.

- Producers need to be informed and realistic as to what should be done.

Listing of Consumer Issues/Concerns

- Diet and health

- Food safety

- Bacterial resistance to antibiotics
- Transmission of disease from livestock to humans
- Residues in foods from livestock
- Product perceptions: some healthy and some not
- Media influence on consumer concerns
- Quality assurance programs to guarantee consumer safety

Diet and Health

- Concerns include cholesterol, saturated fats, sugar
- Individual genetic difference in response to diet
- Smoking contributes to health problems
- Poor choices made by consumers contribute to health problems
- Dietary guidelines provided by the Government for healthy eating
- Dietary habits and lifestyles keep changing and some are fads

Food Safety Concerns

- Microbial contamination of meat such as *E. coli* 0157:H7
- Natural toxicants getting in to the food supply
- Environmental toxicants; for example, mercury (Hg) in fish
- Pesticide residues in animal products
- Food additives in animal products – safe or not

Food Safety Risks in Perspective

- Risks are relative; for example, death from the following:
    - Auto accident = 220 per million

- Drowning = 38 per million
- Fire-related = 29 per million
- Lightning = 0.6 per million
- Venom = 0.2 per million
- Botulism = 0.02 per million
- Salmonellosis = 0.01 per million

Marketing Issues

- Concentration and integration of the livestock industry concerns some producers and consumers.
    - Who controls?
- How do global markets affect consumers and producers?
- More value-based marketing occurring
    - What is the product worth to the consumer? For example, organic or natural vs conventional.

Issues and Opportunities

- Issues create opportunities such as:
    - New markets
    - New jobs
    - New products
- Calls for creativity, sustainability to provide for the well-being of producers

Chapter Resource

Consumption and Consumer Attitudes

- Total meat consumption trends
    - Beef

- ✓ Consumption decreased
- ✓ Due to diet, health issues and economics

o Pork
- ✓ Consumption has remained relatively level, but has highs and lows
- ✓ Influenced by supply and demand economics

o Lamb
- ✓ Consumption is low and decreased
- ✓ Due to high cost and poor availability
- ✓ Quality also a concern; mutton not favored

Chapter Resource

o Veal
- ✓ Consumption decreased
- ✓ Due to animal rights issues and high cost

o Chicken
- ✓ Consumption increased
- ✓ Due to positive health image, economics (cost) and variety

o Turkey
- ✓ Consumption increased
- ✓ Due to positive health image, economics (cost) and variety

o Fish / Seafood
- ✓ Consumption increased
- ✓ Due to positive health image and increased distribution
- ✓ Limited by availability and quality issues

Summary

Each of the major livestock industries face major environmental, perception, regulatory and human health and safety issues. Consumers have their set of concerns. Livestock producers have their concerns and constraints. Animal welfare is on concern to livestock producers and animal protection groups but some groups tend to go to extremes for animal rights and liberation. Livestock producers show concern for the environment through changes in production methods and good stewardship. Some environmental issues are taken to the extreme to the detriment of the producers and the consumers.

Additional Resources

Taylor, R.E. and T.G. Field. 2010. Scientific farm animal production: Introduction to animal science. 10th ed. Upper Saddle River, NJ: Prentice Hall.

Animal Welfare Information Center (AWIC)
http://www.nal.usda.gov/awic/

CAST – Council for Agricultural Science and Technology
http://www.cast-science.org/

Assessment

Take assessment online here: http://tinyurl.com/AnSci-Issues
Download and print the assessment by scanning this QR code or by going to this URL: http://www.tagmydoc.com/Ch53AnSci

Appendix

Conversion Tables for Common Weights and Measures

Common Measures	Conversion Amounts
1 pound	454 grams
2.2 pounds	1 kilogram
1 quart	.9464 liter
1 gram	15.43 grains
1 metric ton	2,205 pounds
1 inch	2.54 centimeters
1 centimeter	10 millimeters or .39 inches
1 meter	39.37 inches
1 acre	.406 hectare

Conversion Factors for English and Metric Measurements

To Convert the English	To the Metric Multiply by	To Convert Metric	Multiply by	To Get English
acres	0.4047	hectares	2.47	acres
acres	4047	square meters	0.000247	acres
BTUs	1055	joules	0.000948	BTUs
BTUs	0.0002928	kilowatt hours	3415.301	BTUs
BTU/hours	0.2931	watts	3.411805	BTU/hours
bushels	0.03524	cubic meters	28.37684	bushels
bushels	35.24	liters	0.028377	bushels
cubic feet	0.02832	cubic meters	35.31073	cubic feet
cubic feet	28.32	liters	0.035311	cubic feet
cubic inches	16.39	cubic centimeters	0.061013	cubic inches
cubic inches	0.01639	liters	61.01281	cubic inches
cubic yards	0.7646	cubic meters	1.307873	cubic yards
cubic yards	764.6	liters	0.001308	cubic yards
feet	30.48	centimeters	0.032808	feet
feet	0.3048	meters	3.28084	feet
feet/minute	0.508	centimeters/second	1.968504	feet/minute
feet/second	30.48	centimeters/second	0.032808	feet/second
gallons	3785	cubic centimeters	0.000264	gallons
gallons	0.003785	cubic meters	264.2008	gallons
gallons	3.785	liters	0.264201	gallons
gallons/minute	0.06308	liters/second	15.85289	gallons/minute
inches	2.54	centimeters	0.393701	inches
inches	0.0254	meters	39.37008	inches
miles	1.609	kilometers	0.621504	miles
miles per hour	26.82	meters/minute	0.037286	miles per hour
ounces	28.349	grams	0.035275	ounces
fluid ounces	0.02947	liters	33.93281	fluid ounces
liquid pints	0.4732	liters	2.113271	liquid pints
pounds	453.59	grams	0.002205	pounds
quarts	0.9463	liters	1.056747	quarts
square feet	0.0929	square meters	10.76426	square feet
square yards	0.8361	square meters	1.196029	square yards
tons	0.9078	tons	1.101564	tons
yards	0.0009144	kilometers	1093.613	yards
yards	0.9144	meters	1.093613	yards

National Agriculture, Food and Natural Resources (AFNR) Career Cluster Content Standards

National AFNR Content Standards, Revised 2015

Source: https://www.ffa.org/thecouncil/afnr

Purpose

The AFNR Career Cluster Content Standards provide state agricultural education leaders and educators with a high-quality, rigorous set of standards to guide what students should know and be able to do after completing a program of study in each of the AFNR career pathways.

State leaders and local educators are encouraged to use the standards as a guide for the development of well-planned curriculum and assessments for AFNR-related CTE programs. These standards are intended to help shape the design of all components of an agricultural education program including:
- Classroom and laboratory instruction.
- Career and Technical Student Organization (CTSO) experiences through organizations such as the National FFA Organization and the Post-Secondary Agriculture Students Organization (PAS).
- Work-based learning experiences such as Supervised Agricultural Experience (SAE) Programs and internships.

Process

The 2015 revision focused on ensuring that the content standards:
- Reflect essential and up-to-date knowledge and skills that students need to be ready for early-career success in a variety of AFNR disciplines.
- Provide a sound basis upon which to design AFNR related Career and Technical Education (CTE) courses.
- Provide a sound basis for developing end of course/program assessments to measure students' attainment of essential disciplinary knowledge and skills.
- The multi-stage review and revision process began in 2014 and was informed by input and guidance from more than 270 secondary and post-secondary educators, business, industry and state leaders in career and technical education

Alignments and Crosswalks

The National Association of State Directors of Career and Technical Education/National Career Technical Education Foundation (NASDCTEc/NCTEF) have provided permission to use the Common Career and Technical Core (CCTC) Standards in support of this project. In addition, the standards have been reviewed to identify crosswalks to the following national academic standards:
- Common Core English Language Arts

- Common Core Mathematics
- Next generation Science Standards
- Green/Sustainability Knowledge and Skill Statements
- National Standards for Financial Literacy
- AFNR Documents AFNR Career Cluster Content Standards
- AFNR Career Cluster Content Standards CROSSWALKS
- AFNR Cluster Skills: Agribusiness Systems Career Pathway; Animal Systems Career Pathway; Biotechnology Systems Career Pathway; Career Ready Practices; Environmental Service Systems Career Pathway; Food Products and Processing Systems Career Pathway; Natural Resource Systems Career Pathway; Plant Systems Career Pathway and Power, Structural and Technical Systems Career Pathway

Availabiltiy

All of the National AFNR Content Standards and associated information are available online at the National FFA website or by following this link: https://www.ffa.org/thecouncil/afnr

Also, the complete document of the National AFNR Content Standard can be downloaded from the National Agricultural Institute's "TagMyDoc" site by following this URL: : http://www.tagmydoc.com/AFNRSTDS

Or by scanning this QR code:

Glossary

A

abdominal cavity - The largest body cavity in humans and many animals and holds the bulk of the viscera.

abnormal behavior - Vices; not normal for the species.

abomasum - Fourth stomach compartment of ruminant animals that corresponds to the true stomach of monogastric animals.

abortion - Unexpected delivery of fetus between conception and a few days before normal parturition.

abscess - Localized collection of pus in a cavity formed by disintegration of tissues.

absorption - The passage of liquid and digested (soluble) food across the gut wall.

absorptive cells - The single layer of epithelial cells lining the surface of the small intestine selectively absorbs food molecules from the gut into the bloodstream.

accessory sex glands - The seminal vesicles, prostate and Cowper's glands in the male. These glands add their secretions to the sperm to form semen.

Acid Detergent Fiber (ADF) - Least digestible plant components, including cellulose and lignin.

acquired immunity - Associated with the presence of antibodies from another immune animal or from exposure to the disease.

active immunity - Acquired through direct contact with the specific disease-causing organism that causes the body to develop antibodies to combat invasion.

acute - Relatively sudden appearance of symptoms (within 24 hours).

adenosine triphosphate (ATP) - Molecule involved in the "energy currency" of the cell; energy is released when a phosphate group is broken off forming ADP (adenosine diphosphate).

adipocyte - A single fat cell

adipose tissue - Fat cells or fat tissue.

afterbirth - The membranes attached to the fetus that are expelled after parturition.

agnostic behavior - Includes fighting or flight and other reactions involving conflict.

AI - See artificial insemination.

air dry - Refers to feeds in equilibrium with air; they would contain approximately 10% water or 90% dry matter.

albumen - The white of an egg.

alimentary canal - Passageway for food and waste products through the body.

alleles - Genes occupying corresponding loci in homologous chromosomes that affect the same hereditary trait but in different ways.

allelomimetic behavior - Doing the same thing. Animals tend to follow the actions of other animals.

allopatric species - Related species which cannot interbreed because of geographical separation.

alveolus (plural alveoli) - A hollow cluster of cells. In the mammary gland, these cells secret milk.

amino acid - Any of a class of 20 molecules that are combined to form proteins in living things.

amnion - A fluid-filled membrane located next to the fetus.

ampulla - The dilated or enlarged upper portion of the vas deferens in bulls, bucks and rams, where sperm are stored for sudden release at ejaculation.

anabolic - A constructive, or "building up," metabolic process.

anaerobic - Able to survive or function where there is no atmospheric oxygen.

analogous - Comparable to.

anaphase - Continuing mitosis, pairs of identical chromosomes separate from one another.

anatomy - Science of animal body structure and the relation of the body parts.

androgen - A male sex hormone, such as testosterone.

anemia - Deficiency of hemoglobin, often accompanied by a reduced number of red blood cells. Usually results from an iron deficiency.

anestrous - Period of time when female is not in estrus; the nonbreeding season.

animal rights - Rights believed to belong to animals to live free from use in medical research, hunting and other services to humans.

animal welfare - Physical and psychological well-being of animals

antemortem - Before death.

anterior - Situated in front of, or toward the front part of, a point of reference; toward the head of an animal.

anterior pituitary - The part of the pituitary gland, located at the base of the brain, that produces several hormones.

anthelmintic - A drug or chemical agent used to kill or remove internal parasites.

anthropomorphism - Attributing human characteristics to animals.

antibiotic - A product produced by living organisms, such as yeast, which destroys or inhibits the growth of other microorganisms, especially bacteria.

antibody - A specific protein molecule that is produced in response to a foreign protein (antigen) that has been introduced into the body.

antigen - A foreign substance that, when introduced into the blood or tissues, causes the formation of antibodies. Antigens may be toxins or native proteins.

anti-inflammatory - An agent that acts to decrease inflammation and associated pain, heat, and swelling.

antiseptic - A chemical agent used on living tissue to control the growth and development of microorganisms.

antiserum - To help the animal's body fight the virus until the animal's system makes its own antibodies.

antitoxin - An antibody that is capable of neutralizing poisons from animal and vegetable sources.

appendicular skeleton - Includes the four legs, used for locomotion and connects with the axial skeleton by muscles.

aquaculture - Raising of aquatic animals or the cultivation of aquatic plants for food.

artery - Vessel through which blood passes from the heart to all parts of the body.

arthropod - Invertebrate animal with exoskeleton, jointed appendage and segmented body.

artificial insemination (AI) - The introduction of semen into the female reproductive tract (usually the cervix or uterus) by a technique other than natural service.

artificial vagina - A device used to collect semen from a male when he mounts in a normal manner to copulate. The male ejaculates into this device, which simulates the vagina of the female in pressure, temperature, and sensation to the penis.

as fed - Refers to feeds that contain their normal amount of moisture.

ascaris - Any of the genus (*Ascaris*) of parasitic roundworms.

aspirate - Pull back on the syringe plunger to be sure the needle is not in a blood vessel.

assimilation - Process of transforming food into living tissue.

ATP - Adenosine triphosphate, a fully charged energy unit.

atrophy - Shrinking or wasting away of a tissue or organ.

autopsy - A postmortem examination in which the body is dissected to deter mine the cause of death.

avian - Refers to birds, including poultry.

axial skeleton - Includes the skull and spinal column.

axion - Elongated fiber of the nerve.

B

bacteria - Constitute a large domain of prokaryotic microorganisms.

Bacterial Crude Protein (BCP) - Protein supplied to an animal by rumen microbes.

bacterin - A suspension of killed or attenuated bacteria for use as a vaccine.

balance sheet - A statement of assets owned and liabilities owed in dollar terms that shows the equity or net worth at a specific point in time (e.g., net worth statement).

balanced ration - Contains more than one feed, so proper quantities of essential nutrients will be provided.

balling gun - Administers large pills for dosing animals.

band - (1) A relatively large group of range sheep; (2) method of identification (e.g., put a band around the leg of a chicken).

barren - Not capable of producing offspring.

barrow - A male swine that was castrated before reaching puberty.

basal metabolism - The chemical changes that occur in an animal's body when the animal is in a thermoneutral environment, resting, and in a postabsorptive state. It is usually determined by measuring oxygen consumption and carbon dioxide production.

base pair - Two nitrogenous bases (adenine and thymine or guanine and cytosine) held together by weak bonds. Two strands of DNA are held together in the shape of a double helix by the bonds between base pairs.

beef - The meat from cattle (bovine species) other than calves (the meat from calves is called **veal).**

beriberi - A disease caused by a deficiency of vitamin B_1.

bilateral symmetry - The body is divided into equivalent right and left halves by only one plane.

binomial nomenclature - A formal system of naming species of living things by giving each a name composed of two parts, both of which use Latin grammatical forms.

biologicals - Medicinal products used primarily to prevent disease, including serums, vaccines, antigens, and antitoxins.

biosecurity - Procedures intended to protect animals against disease or harmful biological agents).

biotechnology - The use of microorganisms, plant cells, animal cells, or parts of cells (such as enzymes) to produce industrially important products or processes.

blemish - Any defect or injury that mars the appearance of, but does not impair the usefulness of, an animal.

bloat - An abnormal condition in ruminants characterized by a distention of the rumen, usually seen on an animal's upper left side, owing to an accumulation of gases.

blood spots - Spots in the egg caused by a rupture of one or more blood vessels in the yolk follicle at the time of ovulation.

boar - A male swine of breeding age.

bog spavin - A soft enlargement of the anterior, inner aspect of the hock.

bolus - (1) Regurgitated food; (2) a large pill for dosing animals.

bone spavin - A bony (hard) enlargement of the inner aspect of the hock.

bots - Any of a number of related flies whose larvae are parasitic in horses and sheep.

bovine - A general family grouping of cattle.

break joint - Denotes the point on a lamb carcass where the foot and pastern are removed at the cartilaginous junction of the front leg.

bred - Female has been mated to the male. Usually implies the female is pregnant.

breech - A presentation at birth is where the rear portion of the fetus is presented first.

breed - Animals of common origin with characteristics that distinguish them from other groups within the same species.

breeder - A person who breeds livestock, racehorses, other animals, or plants.

breeding value - A genetic measure for one trait of an animal, calculated by combining into one number several performance values that have been accumulated on the animal and the animal's relatives.

British thermal unit - (Btu) The quantity of heat required to raise the temperature of 1 lb of water 1°F or near 39.2°F.

broiler - A young meat-type chicken of either sex (usually 6 to 8 weeks of age) weighing 3 to 5 lb. Also referred to as a fryer or young chicken.

brood - Mother

brooder - Fish that have reached reproductive maturity.

broodiness - The desire of a female bird to sit on eggs (incubate).

browse - Woody or brushy plants. Livestock feed on tender shoots or twigs.

brucellosis - A contagious bacterial disease that results in abortions; also called Bang's disease.

BST (Bovine Somatotropin) - Growth hormone produced from pituitary gland of cattle.

buck - A male sheep or goat. This term usually denotes animals of breeding age.

bulbourethral - (Cowper's) gland An accessory gland of the male that secretes a fluid which constitutes a portion of the semen.

bull - A bovine male. The term usually denotes animals of breeding age.

by-product - A product of considerably less value than the major product. For example, in U.S. meat animals, the hide, pelt and offal are by-products, whereas meat is the major product.

C

C-section - See cesarean section.

CAFOs - Confined Animal Feeding Operations

calcification - Normally occurs in the formation of bone, but calcium can be deposited abnormally in soft tissue, causing it to harden.

calf - A young male or female bovine animal under 1 year of age.

calorie - The amount of heat required to raise the temperature of 1 g of water from 15°C to 16°C.

calve - In cattle, giving birth; same as parturition.

calving interval - Amount of time (days or months) between the birth of a calf and the birth of a subsequent calf, both from the same cow.

candling - Shining of a bright light through an egg to see if it contains a live embryo.

canter - A slow, easy gallop.

capacitation - One of the last steps in the maturation of mammalian spermatozoa and is required to render them competent to fertilize an oocyte (egg).

capon - Castrated male chicken. Castration usually occurs between 3 and 4 weeks of age.

carbohydrates - Any foods, including starches, sugars, celluloses and gums, that are broken down to simple sugars through digestion.

cardiac - Pertaining to the heart.

carnivores - Animals that feed on flesh.

carnivorous - Subsisting or feeding on animal tissues.

carotene - The orange pigment found in carrots, leafy plants, yellow corn and other feeds, which can be broken down to form two molecules of vitamin A.

carotenoids - Natural pigments found in plants and animals.

carrying capacity - Number of people, other living organisms, or crops that a region can support without environmental degradation.

casein - Major protein of milk.

cash-flow statement - A financial statement summarizing all cash receipts and disbursements over the period of time covered by the statement.

castration - To remove the testicles.

catalysts - A substance that increases the rate of a chemical reaction without itself undergoing any permanent chemical change.

caudal - At or near the tail or the posterior part of the body.

cecal fermenters - Digest nutrients by means of the cecum rather than by a multi-chambered stomach such as the rabbit and horse.

cecum (ceca) - Large, sock-shaped pouch between the horse's small and large intestines; important in cellulose digestion.

cell - A specific, separate mass of living material that is surrounded by a semi-permeable membrane.

cell wall - (not membrane) Composed of two layers which provide support and protection for the cell.

cellulose - An insoluble substance that is the main constituent of plant cell walls and of vegetable fibers such as cotton.

centriole - See centrosome

centrosome - Near the nucleus and functions in cell division.

cervix - Portion of the female reproductive tract between the vagina and the uterus. It is usually sealed by thick mucus except when the female is in estrus or delivering young.

cesarean section - Delivery of fetus through an incision in abdominal and uterine walls. (See c-section)

chevon - Meat from goats.

chick - A young chicken that has recently been hatched.

choice - Most economical and most desirable carcass grade.

chromosome - Self-replicating genetic structure of cells containing the cellular DNA that bears in its nucleotide sequence the linear array of genes.

chronic - That which develops more slowly, lingers and will frequently reappear.

cilia - A short, microscopic, hair like vibrating structure.

class - A group of animals categorized primarily by sex and age.

clitoris - Ventral part of the vulva of the female reproductive tract that is homologous to the penis in the male. It is highly sensory.

clutch - Eggs laid by a hen on consecutive days.

coccidia - A protozoan organism that causes an intestinal disease called coccidiosis.

coccidiosis - A morbid state caused by the presence of organisms called coccidia, which belong to a class of sporozoans.

coccygeal - Referring to the coccyx, the small tail-like bone at the bottom of the spine.

cock - A male chicken; also called a rooster.

cockerel - Immature male chicken.

cod - Scrotal area of steer remaining after castration.

codons - A triplet (3) of nucleotides bases.

colic - A nonspecific pain of the digestive tract.

collagen - Main structural protein found in animal connective tissue, yielding gelatin when boiled.

colloidal suspension - A mixture in which small particles of a substance are dispersed throughout a gas or liquid.

colon - The large intestine from the end of the ileum and beginning with the cecum to the anus.

colostrum - First milk given by a female after delivery of her young. It is high in antibodies that protect young animals from invading microorganisms.

colt - A young male of the horse or donkey species.

comb - Fleshy outgrowth on the top of a chicken's head, usually red in color, with varying sizes and shapes.

commercial - (1) A carcass grade of cattle; (2) livestock that are not registered or pedigreed by a registry (e.g., breed) association.

commercial herds - Large group of animals kept together as livestock

companion animal - Used to indicate that a human is frequently in the company of an animal.

comparative anatomy - Comparison of parts, organs, etc. of different species.

complete feed - A nutritionally adequate feed for animals specifically formulated to be fed as the sole ration and capable of maintaining life and/or promoting production without any additional substance, except water, being consumed.

composite breed - A breed that has been formed by crossing two or more breeds; also called synthetic breed.

concentrate - A feed used with another to improve the nutritive balance of the total, and intended to be further diluted and mixed to produce a supplement or a complete feed.

conception - Fertilization of the ovum (egg).

conditioning - Growing program for feeder cattle from the time calves are weaned until they enter a feedlot to be finished on a high protein ration; also called backgrounding.

conformation - Physical form of an animal; its shape and arrangement of parts.

contagious disease - Infectious disease; a disease that is transmitted from one animal to another.

contaminants - Presence of a minor and unwanted constituent in a material such manure, grease, blood, yolk, etc.

coronary band (coronet) - Boundary between the top of the hoof wall and the skin at the bottom of the pastern where hoof growth begins.

corpus luteum - A yellowish body in the mammalian ovary. The cells that were follicular cells develop into the corpus luteum, which secretes progesterone. It becomes yellow in color from the yellow lipids that are in the cells.

cortex - An outer layer of tissue immediately below the epidermis of a stem or root.

cotyledon - An area of the placenta that interfaces with the uterine lining to allow nutrients and wastes to pass from the mother to the developing young. Sometimes referred to as button.

cow - A sexually mature, female bovine animal; usually one that has produced a calf.

cow hocked - A condition in which the hocks are close together but the feet stand apart.

cow-calf operation - A management unit that maintains a breeding herd and produces weaned calves.

Cowper's gland - Either of a pair of small glands that open into the urethra at the base of the penis and secrete a constituent of seminal fluid.

cranial - Applied to the front or head of an animal; directional terms are anterior and superior.

creep - An enclosure in which young can enter to obtain feed but larger animals cannot enter; called creep feeding.

creep feeding - Supplemental feeding

crimp - Waves, or kinks, in a wool fiber.

crossbred - An animal produced by crossing two or more breeds.

crossbreeding - Mating animals from genetically diverse groups (i.e., breeds) within a species.

Crude Fiber(CF) - Traditional measure of fiber content in feeds.

Crude Protein(CP) - Measures the nitrogen content of a feedstuff, including both true protein and non-protein nitrogen.

cryptorchidism - Retention of one or both testicles in the abdominal cavity in animals that typically have the testicles hanging in a scrotal sac.

cud - Bolus of feed a ruminant animal regurgitates for further chewing.

cull - To eliminate one or more animals from the breeding herd or flock.

curd - Coagulated milk.

cutability - Fat, lean, and bone composition of meat animals; used interchangeably with yield grade.

cuticle - The outer layer of the wool fiber.

cwt - An abbreviation for hundredweight (100 lb).

cycling - Infers that nonpregnant females have active estrous (heat) cycles.

cytoplasm - Part between the outer cell membrane and the nuclear membrane.

D

dam - Female parent.

daughter cells - Either of the two cells formed when a cell undergoes cell division by mitosis.

Degradable Intake Protein (DIP) - The fraction of the crude protein which is degradable in the rumen and provides nitrogen for rumen microorganisms to synthesize bacterial crude protein (BCP) which is protein supplied to the animal by rumen microbes.

dehorn - To remove the horns from an animal.

dehydrated - Body is lacking water.

demeanor - Outward behavior.

dentrites - A short branched extension of a nerve cell, along which impulses received from other cells at synapses are transmitted to the cell body.

deoxyribonucleic acid (DNA) - A complex double-stranded molecule consisting of deoxyribose (a sugar), phosphoric acid, and four nitrogen bases (a gene is a piece of DNA). This molecule encodes genetic information; it is held together by weak bonds between base pairs of nucleotides.

depreciation - An accounting procedure by which the purchase price of an asset with a useful life of more than 1 year is prorated over time.

dermatitis - Inflammation and redness of skin.

detoxifies - Removes toxic substances or qualities.

dewclaws - Hard horny structures above the hoof on the rear surface of the legs of cattle, swine and sheep.

dewlap - Loose skin under the chin and neck of cattle.

DHIA Dairy Herd Improvement Association - An association which dairy producers participate in keeping dairy records. Sanctioned by the National Cooperative Dairy Herd Improvement Program.

DHIR Dairy Herd Improvement Registry - A dairy record-keeping plan sponsored by the breed associations.

diestrus - A period of sexual inactivity between recurrent periods of estrus.

diet - Feed ingredients or mixture of ingredients (including water), which are consumed by animals.

Diethylstilbestrol (DES) - A synthetic (estrogen) compound recognized by the estrogen receptors as a steroid; no longer used in production.

differentiate - Process by which a less specialized cell becomes a more specialized cell type

diffusion - Movement of a substance from a place where it is found in high concentration (relatively large amounts) to a place of low concentration (relatively small amounts)

digesta - Partially digested food.

digestibility - Quality of being digestible. If a high percentage of a given food taken into the digestive tract is absorbed into the body, that food is said to have high digestibility.

digestible nutrient - That portion of a nutrient which may be broken down (digested) and absorbed and used by the body

Digestible Protein (DP) - Reported by some laboratories but protein digestibility is influenced by external factors.

digestion - The reduction in particle size of feed so that the feed becomes soluble and can pass across the gut wall into the vascular or lymph system.

diploid - Having the normal, paired chromosomes of somatic tissue as produced by the doubling of the primary chromosomes of the germ cells at fertilization.

disease - Any deviation from a normal state of health.

disinfect - To kill, or render ineffective, harmful microorganisms and parasites.

disinfectant - A chemical that destroys disease-producing microorganisms or parasites.

distal - Position that is distant from the point of attachment of an organ.

distended - Swollen

DM - See dry matter.

DNA - See deoxyribonucleic acid.

DNA sequence - The relative order of base pairs, whether in a fragment of DNA, a gene, a chromosome, or an entire genome.

dock - (1) To cut off the tail; (2) the remaining portion of the tail of a sheep that has been docked; (3) to reduce or lower in value.

doe - A female goat or rabbit.

dominance - (1) A situation in which one gene of an allelic pair prevents the phenotypic expression of the other member of the allelic pair; (2) a type of social behavior in which an animal exerts influence over one or more other animals.

dominant gene - A gene that overpowers and prevents the expression of its recessive allele when the two alleles are present in a heterozygous individual.

dorsal - Of, on, or near the back of an animal.

draft horse - Large horses that usually stand taller than 16 hands at the withers

dressing percentage - Percentage of the live animal weight that becomes the carcass weight at slaughter. It is determined by dividing the carcass weight by the liveweight, then multiplying by 100. Used interchangeably with yield.

dry (cow, ewe, sow, mare) - Refers to a nonlactating female.

dry matter (DM) - Feed after water (moisture) has been removed (100% dry).

drylots - No pasture; daily feed and water is provided by the caretaker.

dwarfism - State of being abnormally undersized. Two kinds of dwarfs are recognized: proportionate and disproportionate.

dystocia - Difficult birth.

E

edema - Abnormal collection of fluid in body tissues that causes soft swelling.

ejaculation - Discharge of semen from the male.

eliminative behavior - Involves voiding of feces and urine.

emasculator - Tool used for castration.

embryo - Very early stage of individual development within the uterus. The embryo grows and develops into a fetus. In poultry, the embryo develops within the eggshell.

embryo transfer (ET) - The transfer of fertilized eggs from a donor female to one or more recipient females.

embryology - The study of body before birth.

endocrine gland - A ductless gland that secretes a hormone into the bloodstream.

endocrinology - Science that deals with the study of the endocrine glands and their secretions, the hormones.

endometrium - Mucous membrane that lines the uterus.

endoplasmic reticulum - A network of membranous tubules within the cytoplasm of a eukaryotic cell, continuous with the nuclear membrane. It usually has ribosomes attached and is involved in protein and lipid synthesis.

enterotoxemia - A disease of the intestinal tract caused by bacterial secretion of toxins.

environment - Sum total of all external conditions that affect the well-being and performance of an animal.

enzyme - A complex protein produced by living cells that causes changes in other substances in the cells without being changed itself and without becoming a part of the product.

epididymis - Long, coiled tubule leading from the testis to the vas deferens.

epididymitis - An inflammation of the epididymis.

epimeletic behavior - Caregiving and care-seeking behaviors.

epinephrine - Adrenaline

epiphysis - A piece of bone separated from a long bone in early life by cartilage, which later becomes part of the larger bone.

epistasis - A situation in which a gene or gene pair masks (or controls) the expression of another nonallelic pair of genes.

equine - Refers to horses.

equine encephalomyelitis - An inflammation of the brain of horses.

erythrocytes - Red blood cells.

esophageal groove - A groove in the reticulum between the esophagus and omasum. Directs milk in the nursing young ruminant directly from the esophagus to the omasum.

esophagus - Muscular tube that connects the pharynx to the stomach

essential amino acids - Those which cannot be made in the body from other substances, or which cannot be made in sufficient amounts for physiological (body function) needs.

essential nutrient - A nutrient that cannot be synthesized by the body and must be supplied in the diet.

Estradiol benzoate - A natural type of estrogen (female hormone) combined with the chemical benzoate.

estrogen - Any hormone (including estradiol, estriol, and estrone) that causes the female to come physiologically into heat and to be receptive to the male. Estrogens are produced by the follicle of the ovary and by the placenta.

estrous - An adjective meaning "heat," which modifies such words as cycle. The estrous cycle is the heat cycle, or time from one heat to the next.

estrous synchronization - Controlling the estrous cycle so that a high percentage of the females in the herd express estrus at approximately the same time.

estrus - The period of mating activity in the female mammal. Same as heat.

ET - See embryo transfer.

ethology - The study of animal behavior

eukaryote - Cell or organism with membrane-bound, structurally discrete nucleus and other well-developed subcellular compartments. Eukaryotes include all organisms except viruses, bacteria, and blue-green algae.

euthanized - Put to death humanely.

eviscerate - Removal of the internal organs during the slaughtering process.

evolution - A change in the genetic makeup of a population with time.

ewe - A sexually mature female sheep. A ewe lamb is a female sheep before attaining sexual maturity.

excretion - Expelling of waste products not useful in the animal's body.

exocrine gland - Gland that secretes fluid into a duct.

F

fallopian tubes - Found at the anterior end of each uterine horn.

farrow - To deliver, or give birth to pigs; same as parturition.

fat - Adipose tissue.

fat-soluble vitamins - Any vitamin that is soluble in fats.

FDA - See Food and Drug Administration.

feces - Bowel movements, excrement from the intestinal tract.

feed additive - Ingredient (such as an antibiotic or hormone-like substance) added to a diet to perform a specific role (e.g., to improve gain or feed efficiency).

feed efficiency - (1) The amount of feed required to produce a unit of weight gain or milk; for poultry, this term can also denote the amount of feed required to produce a given quantity of eggs; (2) The amount of gain made per unit of feed.

feed mill - Convert raw materials into finished feed according to very specific formulas developed by nutritionists.

feeder - Animals (e.g., cattle, lambs, pigs) that need further feeding prior to slaughter.

felting - Intermingling of wool fibers

feral - Domesticated animals that return to nature to survive and reproduce.

fermenting - An anaerobic process that converts sugar to acids, gases and/or alcohol.

fertility - The capacity to initiate, sustain, and support reproduction. With reference to poultry, the term typically refers to the percentage of eggs that, when incubated, show some degree of embryonic development.

fertilization - Process in which a sperm unites with an egg to produce a zygote.

fetus - Later stage of individual development within the uterus. Generally, the new individual is regarded as an embryo during the first half of pregnancy, and as a fetus during the last half.

filly - A young female horse.

fingerlings - Young fish, usually 1 to 6 in. long.

finish - Degree of fatness of an animal.

fistula - A running sore at the top of the withers of a horse, resulting from a bruise followed by invasion of microorganisms.

fleece - Wool shorn at one time from all parts of the sheep.

flehmen - A pattern of behavior expressed in some male animals (e.g., bull, ram, stallion) during sexual activity. The upper lip curls up and the animal in hales in the vicinity of the vulva or urine.

floating - Filing horses teeth.

flock - A group of sheep or poultry.

flushing - Placing females (typically sheep and swine) on a gaining level of nutrition before breeding to stimulate greater ovulation rates; also, a behavior in fish whereby diseased fish rub against objects in tanks or ponds.

foal - A young male or female horse (noun) or the act of giving birth (verb).

follicle - A blister-like, fluid-filled structure in the ovary that contains the egg.

follicle-stimulating hormone (FSH) - A hormone produced and released by the anterior pituitary that stimulates the development of the follicle in the ovary.

Food and Drug Administration (FDA) - A U.S. government agency responsible for protecting the public against impure and unsafe foods, drugs, veterinary products, and other products.

food nutrient - A substance that provides nourishment for growth or metabolism.

footrot - A disease of the foot in sheep and cattle. In sheep, it causes rotting of tissue between the horny part of the foot and the soft tissue underneath.

forages - Plant material, leaves and stems.

forb - Weedy or broadleaf plants, as contrasted to grasses, that serve as pasture for animals.

founder - Nutritional ailment resulting from overeating. Lameness in front feet with excessive hoof growth usually occurs.

freemartin - Female born twin to a bull (approximately 9 of 10 will not conceive).

freshen - To give birth to young and initiate milk production. This term is usually used in reference to dairy cattle.

fundus - The part of a hollow organ (such as the uterus or the gallbladder) that is farthest from the opening.

fungi - Large group of spore-producing organisms that includes microorganisms such as yeasts and molds.

furrowing - The cell pinches in on all sides until two daughter cells are formed.

G

gait - The paces of an animal, especially a horse or dog.

gallop - A three-beat gait in which each of the two front feet and both of the hind feet strike the ground at different times.

gametes - Male and female reproductive cells; the sperm and the egg.

gametogenesis - Process by which sperm and eggs are produced.

gastric enzymes - Enzymes that are secreted in the stomach.

gelding - A male horse that has been castrated.

gene - Fundamental physical and functional unit of heredity.

genetic - Of or relating to genes or heredity.

genetic code - Sequence of nucleotides, coded in triplets (codons) along the mRNA, that determines the sequence of amino acids in protein synthesis.

genitalia - Male and female anatomy.

genome - Sum total of a living organism's genetic material. The genome is divided into chromosomes, which contain genes, and genes are made of DNA.

genomics - Study of genes and their function.

genotype - Genetic constitution, or makeup, of an individual. For any pair of alleles, three genotypes (e.g., **AA, Aa,** and **aa**) are possible.

gestation - Time from breeding or conception of a female until she gives birth to her young.

gilt - A young female swine prior to the time that she has produced her first litter.

gizzard - An organ found in the digestive tract of a chicken.

glycogen Storage form of starch in body.

goiter - Enlargement of the thyroid gland, usually caused by iodine-deficient diets.

Golgi apparatus/bodies - Site of accumulation for cells that synthesize and secrete lipids and proteins.

gomer bulls - An intact male that has undergone a penile deviation, penile removal, or vasectomy to render him incapable of physically breeding cows

gonad - Testis of the male; ovary of the female.

gonadotropin - Hormone that stimulates the gonads.

grade - (1) A designation of live or carcass merit (e.g., choice grade); (2) livestock not registered with registry (e.g., breed) association.

grooming behavior - Seen between animals or as they groom themselves.

gross anatomy - That which can be seen with the naked eye.

growth - Increase in protein over its loss in the animal body. Growth occurs by increases in cell numbers, cell size, or both.

growth promotants - Used to help increase the efficiency of animal production by increasing weight gain and product output.

H

half-life - Time it takes for the body to eliminate half of the substance - it is a common measure for use in describing how long substances stay in an animal's body

hand - Used in measuring the height of horses, equivalent to four inches.

handmating - Bringing a female to a male for service (breeding), after which she is removed from the area where the male is located; same as handbreeding.

hank - A measurement of the fineness of wool. A hank is 560 yards of yarn. More hanks of yarn are produced from fine wools than coarse wools.

haploid - Having a single set of unpaired chromosomes.

hardware stomach Reticulum

hatchery - A place where the hatching of fish or poultry eggs is artificially controlled for commercial purposes.

hay - Harvested forage such as alfalfa hay.

heat - See estrus.

heat increment - Increase in heat production after consumption of feed when an animal is in a thermoneutral environment. It includes additional heat generated in fermentation, digestion, and nutrient metabolism.

heaves - A respiratory defect in horses during which the animal has difficulty completing the exhalation of inhaled air.

heifer - A young female bovine cow before the time that she has produced her first calf.

helminths - A parasitic worm; a fluke, tapeworm, or nematode.

hemoglobin - Iron-containing pigment of the red blood cells. It carries oxygen from the lungs to the tissues.

hen - An adult female domestic fowl, such as a chicken or turkey.

herbivorous - Subsisting or feeding on plants.

herd - A group of animals. Used with beef, dairy, or swine.

hereditary - Passing to offspring through genes.

heritability - Traits passed from generation to generation.

hernia - Protrusion of some of the intestine through an opening in the body wall (also commonly called rupture). Two types of hernias, umbilical and scrotal, occur in farm animals.

heterosis - Performance of offspring that is greater than the average of the parents; usually the amount of superiority of the crossbred over the average of the parental breeds. Also referred to as hybrid vigor.

heterozygous - A term designating an individual that possesses unlike genes for a particular trait.

heterozygous genotype - An organism that has both the dominant and the recessive gene.

hinny - Offspring that results from crossing a stallion with a female donkey(jenny).

histology - Study of tissues.

homeostasis - A state of equilibrium, as in an organism or cell, maintained by self-regulating processes.

homogenized - Milk that has had the fat droplets broken into very small particles so that the milk fat stays in suspension in the milk fluids.

homologous - Corresponding in type of structure and derived from a common primitive origin.

homozygous - A term designating an individual whose genes for a particular trait are alike.

hormone - A chemical substance secreted by a ductless gland; usually carried by the bloodstream to other places in the body where it has its specific effect on another organ.

host-specific - Only live in certain types of animals.

hundred-weight - cwt; 100 pounds.

husbandry - Management and care of farm animals.

hybrid vigor - See heterosis.

hydrolyze - Decompose by reacting with water.

hypothalamus - A portion of the brain found in the floor of the third ventricle. It regulates reproduction, hunger and body temperature and has other functions.

hypoxia - A condition resulting from deficient oxygenation of the blood.

I

ileum - Distal portion of the small intestine.

immune response - Generate antibodies to protect against specific diseases.

immunity - Ability of an animal to resist or overcome an infection.

immunoglobulins - Any of a class of proteins present in the serum and cells of the immune system, that function as antibodies.

impaction - Obstructive lodging of food in the intestine.

implant - To graft or insert material to intact tissues.

implantation - A attachment of the fertilized egg to the uterine wall.

imprinting - Learning associated with maturational readiness.

inbreeding - Mating of individuals who are more closely related than the average individuals in a population. Inbreeding increases homozygosity in the population but it does not change gene frequency.

incisor - A front tooth.

incomplete dominance - A form of intermediate inheritance in which one allele for a specific trait is not completely dominant over the other allele.

incubation period - Time between which an egg is placed into an incubator and the young is hatched.

infection - Invasion of the body tissues by microbial agents or parasites other than insects.

infectious - Capable of invading and growing in living tissues; describes various pathogenic microorganisms such as viruses, bacteria, protozoa, and fungi.

ingest - Anything taken into the stomach.

ingesta - Substances taken into the body as nourishment.

ingestive behavior - Includes the mechanics of eating and chewing, obtaining food and water - the daily patterns of feeding.

inheritance - Transmission of genes from parents to offspring.

insemination - Deposition of semen in the female reproductive tract.

instinct - Inborn behavior.

insulin - Hormone secreted by the pancreas to control blood sugar level and utilization of sugar in the body.

integration - Bringing together of all segments of a livestock or poultry production program under one centrally organized unit.

intelligence - Ability to learn to adjust successfully to situations.

interphase - Resting phase between successive mitotic divisions of a cell, or between the first and second divisions of meiosis.

interstitial cells - Cells between the seminiferous tubules of the testicle that produce testosterone.

interstitial fluid - A solution that bathes and surrounds the cells of multicellular animals.

intravenous - Within the vein. An intravenous injection is an injection into a vein.

invertebrates - Animals lacking a backbone.

investigative behavior - Shown when animals explore or investigate a new environment or object.

J

jennet - A female donkey.

jenny - A female donkey.

K

kemp - Coarse, opaque, hairlike fibers in wool.

ketosis - A condition (also called acetonemia) that is characterized by a high concentration of ketone bodies in the body tissues and fluids.

kid - Young goat.

kilocalorie (kcal, Kcal) - An amount of heat equal to 1,000 calories. See also calorie.

kingdom - First and largest division of living things - plants and animals.

kosher meat - Meat from ruminant animals with split hooves where the animals have been slaughtered according to Jewish law.

L

lactalbumin - A nutritive protein of milk.

lactation - Secretion and production of milk.

lactation curve - Period during which the mammary glands secrete milk.

lactoglobulin - A crystalline protein fraction.

lactose - Milk sugar; when digested, it is broken down into one molecule of glucose and one of galactose.

lamb - (1) A young male or female sheep, usually less than 1 year of age; (2) to deliver, or give birth to, a lamb.

lambing - Act of giving birth. Same as parturition.

lambing jug - A small pen in which a ewe is put for lambing. It is also used for containing the ewe and her lamb until the lamb is strong enough to run with other .ewes and lambs.

laminitis - Inflammation of the sensitive plates of soft tissue (laminae) within the horse's foot caused by physical or physiologic injury. Severe cases of laminitis may result in founder, an internal deformity of the foot. Acute laminitis sets in rapidly and usually responds to appropriate, intensive treatment; chronic laminitis is a persistent, long-term condition that may be unresponsive to treatment.

layer - A hen that is kept for egg production.

legume - Any plant of the family *Leguminosae*, such as pea, bean, alfalfa and clover.

lethargic - Sluggish

leukocytes - White blood cells.

LH - See luteinizing hormone.

libido - Sex drive or the desire to mate on the part of the male.

lice - Small, flat, wingless insect with sucking mouthparts that is parasitic on the skin of animals.

ligaments - Strong white fibrous tissues that connect bone to bone.

lipid - An organic substance that is soluble in alcohol or ether but insoluble in water; used interchangeably with the term fat.

litter - The young produced by multiparous females such as swine. The young in a litter are called littermates.

liver flukes - A parasitic flatworm found in the liver.

lobules - A small lobe.

locus - Place on a chromosome where a gene is located.

longevity - Life span of an animal; usually refers to a long life span.

lumbar group - Lower back.

luteinizing hormone (LH) - A protein hormone, produced and released by the anterior pituitary, which stimulates the formation and retention of the corpus luteum. It also initiates ovulation.

lymph - Transparent, nutritive yellow liquid that exudes from blood vessels into tissue spaces and is drained back into the veins through lymph vessels. Lymph playsan important role in fighting infection and maintaining the body's fluid balance.

lysosomes - Small bodies where large numbers of enzymes are stored.

M

macrominerals - Minerals a body needs in larger amounts, includes: calcium, phosphorus, magnesium, sodium, potassium, chloride and sulfur.

maintenance - A condition in which the body is maintained without an increase or decrease in body weight and with no production or work being done.

mammary gland - Gland that secretes milk.

mandible - Lower jaw or jawbone.

marbling - Distribution of fat in muscular tissue; intramuscular fat.

marbling scores - Amount of fat interspersed in the muscle.

mare - A sexually developed female horse.

marrow - Soft center of the bone.

masticate - To chew food.

mastitis - Inflammation of the mammary gland.

masturbation - Ejaculation by a male by some process other than sexual intercourse.

maturity scores - Reflects age of animal at slaughter.

mean - (1) Statistical term for average; (2) term to describe animals having bad behavior.

medulla - Inner region of an organ or tissue, especially when it is distinguishable from the outer region or cortex.

meiosis - A special type of cell nuclear division that is undergone in the production of gametes (sperm in the male, ova in the female). As a result of meiosis, each gamete carries half the number of chromosomes of a typical body cell in that species.

melengestrol acetate (MGA) - A feed additive that suppresses estrus in heifers and is widely used in the feedlot industry.

messenger RNA (mRNA) - RNA that serves as a template for protein synthesis.

metacarpal bones - Extend from the knee to fetlock.

metabolism - (1) The sum total of chemical changes in the body, including the "building up" and "breaking down" processes; (2) the transformation by which energy is made available for body uses.

Metabolizable Protein (MP) - Protein available to the animal including microbial protein synthesized by the rumen microorganisms.

metacarpal bones - Extend from the knee to fetlock.

metaphase - Second stage of cell division, between prophase and anaphase, during which the chromosomes become attached to the spindle fibers.

metestrus - Period immediately following estrus.

metritis - Inflammation (infection) of the uterus.

MGA - See melengestrol acetate.

microclimate - A small, special climate within a macroclimate created by the use of such devices as shelters, heat lamps and bedding.

micromineral - A mineral that is needed in the diet in relatively small amounts. The quantity needed is so small that such a mineral is often called a trace mineral; for example: iron, iodine, zinc and selenium.

micronutrients - Required in small amounts.

milk fat - Fat in milk; synonymous with butterfat.

milk letdown - Release of milk into the teat cisterns.

milk-ejection reflex - An example of endocrine gland activity.

mimicry - Animals simply doing what the other animals in the herd or group are doing.

mites - Very small arachnids that are often parasitic upon animals.

mitochondria - An organelle found in large numbers in most cells, in which the biochemical processes of respiration and energy production occur.

mitosis - A process in which a cell divides to produce two daughter cells, each of which contains the same chromosome complement as the mother cell from which they came.

Modified Live Viruses(MLV) - Products which contain a live virus but have been changed or modified so as to not cause the disease but still stimulate antibody formation against the disease.

mohair - Fleece of the Angora goat.

monogastric - Having only one stomach or only one compartment in the stomach. Examples are swine and poultry.

monorchid - A male with one fertile testicle.

morbidity - Measurement of illness; morbidity rate is the number of individuals in a group that become ill during a specified time.

mortality - State of being subject to death.

motility - Ability to move under their own power.

mottled - Spotted or blotched.

mouth - Initial opening of the alimentary canal.

mucous membranes - An epithelial tissue that secretes mucus which lines many body cavities and tubular organs including the gut and respiratory passages.

mule - Hybrid that is produced by mating a male donkey with a female horse. They are usually sterile.

mutation - A change in a gene.

mutton - Meat from a sheep that is over 1 year old.

mutualism - Specific type of symbiosis between man and animal.

muzzle - Nose of horse, cattle, or sheep.

myofibrils - Primary component part of muscle fibers.

N

nasal cavity - Cavity in which the olfactory organs of vertebrate animals are located.

natural breeds - Selected by human preference or regional diversity.

natural immunity - Refers to the protection an animal has when it is born.

natural selection - Sequence of events that lead to a certain characteristic being selected by the environment.

navel - Area where the umbilical cord was formerly attached to the body of the offspring.

necropsy - Perform a postmortem (after death) examination.

net energy - Metabolizable energy minus heat increments; the energy available to the animal for maintenance and production.

neuron - A nerve cell which transmits messages from one part of the body to another.

Neutral Detergent Fiber(NDF) - Useful measures of feeding value, and should be used to evaluate forages and formulate rations.

nipple - See teat.

Nitrogen-Free Extract (NFE) - Represents carbohydrates, sugars, starches and a major portion of materials classed as hemicellulose in feeds.

nitrogenous - Contains the element nitrogen.

nomenclature - Giving and using of names.

Nonprotein Nitrogen (NPN) - Nitrogen in feeds from substances such as urea and amino acids, but not from preformed proteins.

nonruminant - Simple-stomached or monogastric animal.

NPN - See nonprotein nitrogen.

nucleotide - Subunit of DNA composed of a five-carbon sugar, a nitrogenous base, and a phosphate group.

nucleus - Contains the hereditary, genetic, information and is the control center for the cell.

nutrient - (1) A substance that nourishes the metabolic processes of the body; (2) the end product of digestion.

nutrition Science dealing with the utilization of feed/food by the body and all body processes which transform feed/food into body tissues and activities.

nutrient density - Amount of essential nutrients relative to the number of calories in a given amount of food.

O

offal - All organs and tissues removed from inside the animal during the slaughtering process.

omasum - One of the stomach components of ruminant animals that has many folds.

omnivorous - Feeding on both animal and vegetable substances.

oocyte - Ovulated while in the metaphase of meiosis II.

oogenesis - Process by which eggs, or ova, are produced.

organelle - A structure or part that is enclosed within its own membrane inside a cell and has a particular function.

osmosis - Passage (diffusion) of water across a membrane as a result of different concentrations on the two sides of the membrane; movement of water from area of higher concentration to area of lower concentration.

osteoblasts - Cells that form layers of bone in the early stages of ossification (bone formation).

ova - Plural of ovum, meaning eggs.

ovary - Female reproductive gland in which the eggs are formed and progesterone and estrogenic hormones are produced.

oviduct - A duct leading from the ovary to the horn of the uterus.

ovine - Refers to sheep.

ovulation - Shedding, or release, of the egg from the follicle of the ovary.

ovum - Egg produced by a female.

oxytocin - A hormone released by the pituitary gland that causes increased contraction of the uterus during labor and stimulates the ejection of milk into the ducts of the mammary glands

P

pace - A lateral two-beat gait in which the right rear and front feet hit the ground at one time and the left rear and front feet strike the ground at another time.

paired structures - Similar right and left structures.

palpation - Feeling by hand.

papillae - Any small, nipple-like process or projection.

parasite - An organism that lives a part of its life cycle in or on, and at the expense of, another organism. Parasites of farm animals live at the expense of the farm animals.

parent cell - A cell that is the source of other cells.

parrot mouth - Upper jaw is longer than lower jaw; also called overshot jaw.

parturition - Process of giving birth.

passive immunity - Acquired by transferring of antibodies from an immunized animal to an unimmunized one.

pasteurization - Process of heating milk to 161°F and holding it at that temperature for 15 seconds to destroy pathogenic microorganisms.

pasture rotation - Moving of animals from one pasture to another so that some pasture areas have no livestock on them in certain periods.

pathogen - Biologic agent (i.e., bacteria, virus, protozoa, nematode) that may produce disease or illness.

pathogenic - Infectious agents causing disease.

Pearson square - Helps to formulate seed rations.

pedigree - Record of the ancestry of an animal.

pellets - A small, condensed formed feed.

pelt - Natural, whole-skin covering, including the wool, hair, or fur (e.g., a sheep pelt has the wool left on).

pelvic cavity - Contains the terminal part of the digestive system and all of the internal portions of the urogenital system not in the abdominal cavity.

pen mating - A cohort of females is brought into the boar's pen and he services them all while they are in the pen.

penis - Male organ of copulation. It serves both as a channel for passage of urine from the bladder as an extension of the urethra, and as a copulatory organ through which sperm are deposited into the female reproductive tract.

per capita - Per person.

performance test - Evaluation of an animal according to its performance.

pericardium - A double-walled sac containing the heart and the roots of the great vessels.

peristaltic movement - Muscular contractions that move food through the intestines.

pet - A domestic or tamed animal or bird kept for companionship or pleasure and treated with care and affection.

phalanges - Corresponds to the hand of humans.

pharynx - A short, funnel shaped muscular sac between the mouth and esophagus.

phenotype - Characteristics of an animal that can be seen and/or measured (e.g., the presence or absence of horns, the color, or the weight of an animal).

pheromones - Chemical substances that attract the opposite sex.

photoperiod - Time during which light is present.

phylum - Each new group within a Kingdom

physiology - Science that pertains to the functions of organs, organ systems, or the entire animal.

pin bones - In cattle, the posterior ends of the pelvic bones that appear as two raised areas on either side of the tail head.

pituitary - Small endocrine gland located at the base of the brain.

placenta - Vascular organ that unites the fetus to the uterus.

plant code - A code is printed on every carton produced in a processing plant.

poikilothermic - Having body temperature that varies with the environment.

polar bodies - Minute cell produced and ultimately discarded in the development of an oocyte.

poll evil - An abscess behind the ears of a horse.

polled - Naturally or genetically hornless.

polytocous - Giving birth to several offspring at one time.

Porcine Stress Syndrome (PSS) - A genetic defect in swine inherited as a simple recessive. It is associated with heavily muscled animals that may suddenly die when exposed to stressful conditions. Their muscle is usually pale, soft and exudative (PSE).

posterior - Toward the rear end of an animal.

postgastric fermentation - Fermentation of feed that occurs in the cecum, behind the area where digestion has occurred.

postnatal - See postpartum.

postpartum - After birth.

postpartum interval - Length of time from parturition to when the dam is again pregnant.

poult - A young turkey of either sex, from hatching to approximately 10 weeks of age.

poultry - Term that includes chickens, turkeys, geese, pigeons, peafowls, guineas and game birds.

PQA-Plus - Pork Quality Assurance Plus

predisposing - Inclined to.

pregastric fermentation - Occurs in the rumen of ruminant animals, before feed passes into the portion of the digestive tract in which digestion actually occurs.

pregnancy testing - Evaluation of females for pregnancy through palpation or using an ultrasound machine.

premix - A uniform mixture of one or more micro ingredients with diluent and/or carrier. Premixers are used to facilitate uniform dispersion of the micro ingredients in a large mix.

prenatal - Prior to being born; before birth.

primary breeder - Responsibility is to develop and reproduce strains of chicken that meet the requirements of chicken producer/processing companies.

Prime - Superior marbling, proper carcass conformation and adequate maturity.

probe - A device used to measure backfat thickness in pigs and cattle.

proestrus - Phase of the estrous cycle just before heat (estrus).

progeny testing - An evaluation of an animal on the basis of performance of its offspring.

progesterone - A hormone produced by the corpus luteum that stimulates progestational proliferation in the uterus of the female.

prokaryote - Cell or organism lacking a membrane-bound, structurally discrete nucleus and other subcellular compartments. Bacteria are prokaryotes.

prolapsed - Turned inside out.

pronuclei - Either of a pair of gametic nuclei, in the stage following meiosis but before their fusion leads to the formation of the nucleus of the zygote.

prophase - First stage of cell division, before metaphase, during which the chromosomes become visible as paired chromatids and the nuclear envelope disappears.

prostaglandins - Chemical mediators that control many physiological and biochemical functions in the body. One prostaglandin (PGF_2 alpha) can be used to synchronize estrus.

prostate - A gland of the male reproductive tract that is located just back of the bladder. It secretes a fluid that becomes part of semen at ejaculation.

protein - A large molecule of one or more chains of amino acids in a specific order, the order is determined by the base sequence of nucleotides in the gene coding for the protein.

protein supplement - Any dietary component containing a high concentration(at least 25%) of protein.

protoplasm - Viscid or semi-liquid and jello-like substance which makes up the living cell.

protozoa One-celled, mobile organisms with a nucleus.

proventriculus - Acts as the true stomach of a bird.

proximal - Nearest; the position that is closest to the point of attachment for a limb or bone.

PSE - See pale, soft, exudative.

Psitticosis (parrot fever) - Acute or chronic disease characterized by respiratory and systemic infection.

PSS - See Porcine Stress Syndrome

ptyalin - A form of amylase found in the saliva of humans and some other animals.

puberty - Age at which the reproductive organs become functionally operative.

pullet - Young female chicken from day of hatch through onset of egg production; sometimes the term is used through the first laying year.

purebred - An animal eligible for registry with a recognized breed association.

pylorus - Opening from the stomach into the duodenum (small intestine).

Q

qualitative trait - A trait expressed categorically because of a sharp distinction between phenotypes (e.g., black and red). Usually only one or a few pairs of genes are involved in the expression of a qualitative trait.

quality grades - Animals grouped according to value as prime, choice, etc., based on conformation and fatness of the animals.

quantitative trait - A trait expressed on a continuous/numerical scale because of a gradual variation from one phenotype to another (e.g., weaning weight). Usually many gene pairs and environmental influences are involved in the expression of such traits.

R

rabies - A disease-causing virus transmitted through bites.

race - Considered simply a subdivision of a species which breeds true except for minor variations.

rack - (1) A rapid four-beat gait of a horse; (2) a wholesale cut of lamb located between the shoulder and loin.

ram - A male sheep that is sexually mature.

ram power - Number of rams/number of ewes.

rancid - Spoiled

ration - Amount of total feed fed to an animal over a 24-hour period.

reactive behavior - An animal reacting to its surroundings such as communicating and visual contact with the rest of the herd, a reflex to pain and discomfort or seeking shelter.

receptor cells - Other cells that will respond to a hormone in a target gland or organ.

recessive gene - A gene that has its phenotype masked by its dominant allele when the two genes are both present in an individual.

recombinant DNA (rDNA) - Isolated DNA molecules that can be inserted into the DNA of another cell. rDNA is used in the genetic engineering process.

registered - Recorded in the herd book of a breed.

regurgitate - To cast up digested food to the mouth as is done by ruminants.

reproduction - Production of live, normal offspring.

reproductive glands - Include the testes and ovaries; produce germ or "sex" cells for reproduction and the hormones testosterone and progesterone.

retained placenta - Placenta remains within the reproductive tract after parturition has occurred.

reticulum - One of the stomach components of ruminant animals that is lined with small compartments, giving a honeycomb appearance.

rhinopneumonitis - Equine herpesvirus-1; it produces acute mucus upon primary infection.

ribonucleic acid (RNA) - An essential chemical component of living cells, composed of long chains of phosphate, ribose sugar, and several bases; found in the nucleus and cytoplasm of cells and plays an important role in protein synthesis and other chemical activities of the cell.

ribosomes - A minute particle consisting of RNA and associated proteins, found in large numbers in the cytoplasm of living cells.

rickets - A disease of disturbed ossification of the bones caused by a lack of vitamin D or unbalanced calcium/phosphorus ratio.

ringbone - An ossification of the lateral cartilage of the foot of a horse all around the foot.

riparian - An area next to water (stream, river, or lake) where more vegetation grows (compared to a greater distance from the water source) because of the added moisture from the water. Grazing animals usually inhabit this area more frequently than others, thus increasing the possibility of overgrazing.

RNA - See ribonucleic acid.

roughage - A feed that is high in fiber, low in digestible nutrients, and low in energy. Such feeds as hay, straw, silage and pasture are examples.

rumen - The large fermentation pouch of the ruminant animal in which bacteria and protozoa break down fibrous plant material that is swallowed by the animal; sometimes referred to as the paunch.

ruminant - A mammal whose stomach has four parts (rumen, reticulum, omasum, and abomasum). Cattle, sheep, goats, deer, and elk are ruminants.

rumination - Regurgitation of undigested food and chewing it a second time, after which it is again swallowed.

S

salivary glands - Exocrine glands that secrete juices in the mouth that are mixed with the food.

salmonella - Gram-positive, rod-shaped bacteria that cause various diseases such as food poisoning in animals.

sanitation - Cleanliness

satellite farms - Production facilities located at a different location from the processing facility.

scale - (1) Size; (2) equipment on which an animal is weighed.

scrotum - A pouch that contains the testes. It is also a thermoregulatory organ that contracts when cold and relaxes when warm, thus tending to keep the testes at a lower temperature than that of the body.

scurvy - Swollen and painful joints and bleeding gums in humans and brittleness of bones.

secondary sex characteristics - Those that begin to show with the onset of puberty.

secretion - Production of substances useful for the cells in other parts of the body.

secretory cells - Produce products that are subsequently deposited in either the blood stream or a special duct to an organ, where they are used.

seedstock herds - Breeding cattle typically registered with a breed association.

selection - Differentially reproducing what one wants in a herd or flock.

semen - Fluid containing the sperm that is ejaculated by the male. Secretions from the seminal vesicles, the prostate gland, the bulbourethral glands, and the urethral glands provide most of the fluid.

seminal vesicles - Accessory sex glands of the male that provide a portion of the fluid of semen.

seminiferous tubules - Minute tubules in the testicles in which sperm are produced. They comprise about 90% of the mass of the testes.

sequencing - Putting the amino acids in correct order; determining genetic make-up.

Sertoli cells - Serve a protective and nutritional role for the germ cells (spermatogonia or sex cells).

service - To breed or mate.

settle - To become pregnant.

sex-linked inheritance - Phenotypic expression of an allele related to the chromosomal sex of the individual.

sexual behavior - Involves the courtship, mating and maternal behavior and is controlled by hormones but may be learned.

shearing - Process of removing the fleece (wool) from a sheep.

shoat - A young pig of either sex.

shoe boil - Blemish of the horse caused by the horseshoe putting pressure on the elbow when the horse lies down.

shrink - Loss of weight, commonly used in the loss in live weight when animals are marketed or loss in weight from grease wool to clean wool.

sigmoid flexure - S-shape of the retracted penis in livestock.

silage - Forage, corn fodder, or sorghum preserved by fermentation that produces acids similar to the acids that are used to make pickled foods for people.

simple stomach - Extensive intestinal system with an enlarged cecum.

sinuses - Hollow walled spaces.

sire - Male parent.

skins - Hides from smaller animals such as pigs, sheep, goats and wild animals; a beef hide weighs less than 30 lb.

smooth muscle cells - Spindle-shaped cells that are not striated; they contain one centrally located nucleus per cell.

SNF - See solids-nonfat.

solids-nonfat - Total milk solids minus fat, includes protein, lactose and minerals.

somatic - Body cells.

somatic cell count - An indicator of the quality of milk.

somatotropin - Growth hormone from the anterior pituitary that stimulates nitrogen retention and growth.

sound - A horse who has no lameness or illness.

sow - A female swine that has farrowed one litter or has reached 12 months of age.

spawn - Act of fish laying eggs.

spay - To remove the ovaries.

spermatid - Haploid germ cell prior to spermiogenesis, the formation of sperm.

spermatogenesis - Process by which spermatozoa are formed.

spermatogonia - Sperm producing cells.

spermatozoa - Viable male sex cells.

spermiogenesis - Process by which the spermatid loses most of its cytoplasm and develops a tail to become a mature sperm.

spontaneous mutations - New breeds that showcase a mutation.

stags - Castrated male sheep, cattle, goats, or swine that have reached sexual maturity prior to castration.

stallion - A sexually mature male horse.

standard grade - Usually older animals and thin animals.

staple length - Length of wool fibers.

steer - A castrated bovine male that was castrated early in life before puberty.

sterile - Inability to produce offspring.

steroid - Artificially produced drug similar to the natural hormone that controls inflammation and regulates water balance.

stocker (cattle) - Weaned cattle that are fed high-roughage diets (including grazing) before going into the feedlot.

stocker (fish) - Usually 6 to 12 inches in length and less than 0.75 lb.

strangles - An infectious disease of horses, characterized by inflammation of the mucous membranes of the respiratory tract.

streptococcus - Spherical, gram-positive bacteria that divide in only one plane and occur in chains. Some species cause serious disease.

stress - An unusual or abnormal influence causing a change in an animal's function, structure, or behavior.

striated muscle cells - Voluntary muscle cells that produce movement; connected to bones and contraction causes movement.

strongyles - Any of various roundworms living as parasites, especially in domestic animals.

stud - Usually the same as stallion. Also a place where male animals are maintained (i.e., bull stud).

subcutaneous - Situated beneath, or occurring beneath, the skin. A subcutaneous injection is an injection made under the skin.

subspecies - A subdivision or smaller part of a group of animals (those in a species).

superior - Above or over.

superovulation - Hormonally induced ovulation of a greater than normal number of eggs.

supplement - A feed used with another to improve the nutritive balance of performance of the total and intended to be (1) fed undiluted as a supplement to other feeds, (2) offered free choice with other parts of the ration separately available, or (3) further diluted and mixed to produce a complete feed.

surfactants - Ingredients that reduce surface tensions of liquids and are used to reduce and stop foaming to prevent bloat.

symbiosis - A biological situation in which at least two different kinds of organisms interact; these can include plants, animals, or plant and animal.

sympatric species - Those which can interbreed, but in practice do not because of differences in behavior, breeding, food sources, etc.,

Synovex H - Synthetic steroid used in meat production.

synovial fluid - Secreted by the synovial membrane; helps lubricate the joint.

synthetic - Made by chemical synthesis, especially to imitate a natural product.

synthesized - Formed

T

tack - Equipment used for riding or driving horses.

tags - (1) Wool covered with manure; (2) abbreviated form of ear tags, used for identification.

tail docking - Intentional removal of part of an animal's tail.

taming - On the path to domestication, but a tamed animal is not a domestic animal.

taxonomy - Science of classification and the arrangement of plants and animals into groups based on their natural relationships.

TDN - See Total Digestible Nutrients.

teat - Protuberance of the udder through which milk is drawn.

telophase - Final phase of cell division, between anaphase and interphase, in which the chromatids or chromosomes move to opposite ends of the cell and two nuclei are formed.

temporal bone - Either of a pair of compound bones forming the sides and base of the skull.

tendon - Tough, fibrous connective tissue at ends of muscle bundles that attach muscle to bones or cartilage structures.

testicle - Male sex gland that produces sperm and testosterone.

testosterone - Male sex hormone that stimulates the accessory sex glands, causes the male sex drive, and causes the development of masculine characteristics.

testosterone propionate - A natural hormone which has been combined with a chemical, propionic acid, to increase its half-life.

tetanus - Rigid paralytic disease caused by *Clostridium tetani,* an anaerobic bacterium that lives in soil and feces.

tetrad - A group of four similar chromatids formed by the splitting longitudinally of a pair of homologous chromosomes during meiotic prophase.

thermal - Temperature

thoracic cavity - Chest cavity

thoracic limbs - Arms or front legs (including the scapula, arm, radius, ulna, manus, carpus and digits)

thrombocytes - Platelets in the blood.

thrush - Foot disease characterized by degeneration of the hoof frog and a thick, foul-smelling discharge.

thyroid gland - Two-lobed endocrine gland in the neck that controls the rate at which basic body functions proceed.

thyroxine - Main hormone produced by the thyroid gland, acting to increase metabolic rate and so regulating growth and development.

tibia - Corresponds with the shin bone of humans.

toeing in - Toes of front feet turn in; also called pigeon-toed.

toeing out - Toes of front feet turn out; also called splayfooted.

tom - A male turkey.

tongue - A tool of prehension that is used to grasp the food or to guide it in the mouth and on to the throat.

Total Digestible Nutrients (TDN) - Includes the total amounts of digestible protein, nitrogen-free extract, fiber, and fat (multiplied by 2.25), all summed together.

toxoids - An inactivated, altered toxins (the poison that is produced by pathogenic bacteria) used to stimulate immunity.

trot - A diagonal two-beat gait in which the right front and left rear feet strike the ground in unison, and the left front and right rear strike the ground in unison.

tuberculosis - Caused by bacteria and usually settles in the lungs.

turbidity - Muddiness created by stirring up sediment or having foreign particles suspended.

turbinates - Cartilaginous bone (not hard) covered by highly vascular (many blood vessels) mucosa which serves to clean and warm the air as the animal breathes in.

U

udder - Encased group of mammary glands of animals.

Undegradable Intake Protein (UIP) - Commonly called "bypass protein" because it bypasses rumen breakdown and is mainly digested in the small intestine.

undershot jaw - Lower jaw is longer than upper jaw.

unsoundness - Any defect or injury that interferes with the usefulness of an animal.

urea - Often used as a protein substitute in ruminants. It is a source of nitrogen which the rumen "bugs" can use to make bacterial protein.

urogenital system - Refers to the urinary tract and the accompanying genitalia (male and female anatomy).

uterine horns - Two branches of the uterus.

uterus - That portion of the female reproductive tract where the young develop during pregnancy.

V

vaccination - Act of administering a vaccine or antigens.

vaccine - Suspension of attenuated or killed microbes or toxins administered to induce active immunity.

vacuoles - Storage bodies for water, minerals, etc. and can be large in plant cells.

vagina - Copulatory portion of the female's reproductive tract. The vestibule portion serves for passage of urine during urination; also serves as a canal through which young pass when born.

variation - Deviation from the normal biological form, function, or structure.

vas deferens - Duct that carries sperm from the epididymis to the urethra.

vasectomy - Removal of a portion of the vas deferens. As a result, sperm are prevented from traveling from the testicles to become part of the semen.

veal - Meat from very young cattle, under 3 months of age.

vein - Vessel through which blood passes from various organs or parts back to the heart.

ventral - Lower or abdominal surface of an animal.

ventral cavity - Contains most of the viscera or guts.

vertebrate - Having a spinal column.

vertical integration - A style of business management that allows for maximum control of the products produced.

villi - Projections of the inner lining of the small intestine.

virus - Ultramicroscopic bundle of genetic material capable of multiplying only in living cells. Viruses cause a wide range of disease in plants, animals, and humans, such as rabies and measles.

viscera - Internal organs and glands contained in the thoracic and abdominal cavities.

vitamin - An organic catalyst, or component thereof, that facilitates specific and necessary functions; for example: the B-vitamins, vitamins A, D, E and K.

vulva - External genitalia of a female mammal.

W

walk - A four-beat gait of a horse in which each foot strikes the ground at a time different from each of the other three feet.

water-soluble vitamins - Carried to the body's tissues but are not stored in the body; the B-vitamins and vitamin C.

weaner - An animal that has been weaned or is nearing weaning age.

weaning - Separating young animals from their dams so that the offspring can no longer suckle.

wether - A male sheep castrated before reaching puberty.

white cells (leukocytes, white blood cells) - Colorless blood cells active in the body's defense against infection or other assault; five types are neutrophils, lymphocytes, eosinophils, monocytes and basophils.

white muscle disease - A muscular disease caused by a deficiency of selenium or vitamin E.

withers - Top of the shoulders.

wool - Fibers that grow from the skin of sheep.

Wool Act - The Wool Act of 1699 is an Act of the Parliament of England which attempted to heighten taxation and increase control over colonial trade and production.

wool staple - A cluster of wool fibers.

woolen - A type of yarn that is created from carded wool. It is light, soft and stretchy. It can be used to make blankets, hosiery and flannels.

worsted wool - Made from the long fibers that have been combed to make sure the fibers run the same direction and not carded but washed.

X-Y-Z

yearling - Animals that are approximately 1 year old.

yield - See dressing percentage.

yield grade - Grouping of animals according to the estimated trimmed lean meat that their carcass would provide; used interchangeably with cutability.

yolk - (1) The yellow part of the egg; (2) the natural grease (lanolin) of wool.

yolk sac - Layer of tissue encompassing the yolk of an egg.

Zeranol (Ralgro) - A synthetic compound (not a steroid) and is recognized in the target cells as estrogen.

Zona pellucida - A protective covering around the ova, egg.

zygote - Cell formed by the union of two gametes.